MASTERING THE GRADE 5 TAKS IN SCIENCE

MARK JARRETT
Ph.D., Stanford University

STUART ZIMMER

JAMES KILLORAN

JARRETT PUBLISHING COMPANY

EAST COAST OFFICE
P.O. Box 1460
Ronkonkoma, NY 11779
631-981-4248

SOUTHERN OFFICE
50 Nettles Boulevard
Jensen Beach, FL 34957
800-859-7679

WEST COAST OFFICE
10 Folin Lane
Lafayette, CA 94549
925-906-9742

www.jarrettpub.com
1-800-859-7679   Fax: 631-588-4722

ISBN 1-882422-95-3   [978-1-882422-9-51]
Printed in the United States of America
Second Edition
10 9 8 7 6 5 4 3 2 1   12 11 10 09

ACKNOWLEDGMENTS

The authors would like to thank the following Texas educators who helped review the manuscript. Their collective comments, suggestions, and recommendations have proved invaluable in preparing this book.

Victor Cantu
Grade 5 Science Teacher
Jefferson Elementary School
Edinburg C.I.S.D.
Edinburg, Texas

Carolyn Monmouth
Houston Urban Learning Initiative
in a Network Community
Master Science Lead Teacher
J. Will Jones Elementary School
Houston I.S.D.
Houston, Texas

Dennis Monmouth
Houston Urban Learning Initiative
in a Network Community
Master Science Lead Teacher
Almeda Elementary School
Houston I.S.D.
Houston, Texas

Layout, graphics, and typesetting: Burmar Technical Corporation, Albertson, N.Y.

This book is dedicated...

to my wife, Gośka, and my children Alexander and Julia   — *Mark Jarrett*

to my wife Joan, my children Todd and Ronald, and
my grandchildren Jared and Katie   — *Stuart Zimmer*

to my wife Donna, my children Christian, Carrie, and Jesse,
and my grandchildren Aiden, Christian, and Olivia   — *James Killoran*

TABLE OF CONTENTS

HOW THIS BOOK CAN HELP YOU

Everyone wants to get a high score on the **Elementary Science TAKS**. Unfortunately, just wanting a high score is not enough. You will really have to work at it. With this book as your guide, you should be much better prepared for the test — and even enjoy studying for it. This book provides a complete "refresher" of the knowledge and skills you will need to do your best on the **Elementary Science TAKS**.

Students working together on a class experiment while employing safety practices.

Science Photo Library

This book consists of four units that review what you have learned in science. Each chapter in the book —

★ opens with a list of *Major Ideas* highlighting the most important information.

★ is divided into smaller sections to help you understand major science topics.

★ includes *Applying What You Have Learned* activities to help you practice what you have just learned.

★ includes **Study Cards** to help you review the most important *terms*, *concepts*, and *relationships* described in the chapter.

★ finishes with a *What You Should Know* feature, summarizing the major ideas in the chapter that are often the focus of TAKS questions.

★ provides TAKS-style practice questions at the end; each question is identified by its TAKS objective, grade level, and knowledge-and-skill statement.

You will also find a checklist of the **Elementary Science TAKS** objectives at the end of each unit. Make sure you have mastered each objective before moving to the next unit.

The last part of the book consists of a *complete* practice test, just like the actual **Elementary Science TAKS**.

WHAT IS THE ELEMENTARY SCIENCE TAKS

As a student in Texas, you have probably taken TAKS tests: in reading, writing, and mathematics. This year, you will take your first TAKS test in science. This TAKS will test what you have learned about science in grades 2 through 5.

The **Elementary Science TAKS** has 50 questions, most of which are multiple-choice. These ask a question and give four possible choices for answering it. A few questions on the TAKS may ask you to measure something or to make a simple calculation. To answer these questions you may have to write a number and bubble your answer on a grid-form.

Of the 50 test questions, only 40 will actually count. Ten questions will be sample questions the testmakers are trying out. Unfortunately, you won't know which questions count, so do your best on all questions! Do not leave any questions blank. There's no penalty for guessing a wrong answer. Be sure you answer every question. Take your best guess even when you are unsure of the answer.

The 40 questions on the test that do count will be divided into the following objectives:

Objective	Questions
Objective 1: The Nature of Science These questions test your understanding of scientific investigation, including laboratory safety, measuring, and reading graphs and charts.	13
Objective 2: Life Sciences These questions test your understanding of living things, such as plants, animals, ecosystems, and inherited traits.	9
Objective 3: Physical Sciences These questions test your understanding of motion, force, types of energy, matter, and mixtures.	9
Objective 4: Earth and Space Sciences These questions test your understanding of Earth's processes: weather, the water cycle, erosion, the formation of Earth's resources. They also test your knowledge of the sun, moon, and solar system.	9

As you can see, Objective 1 will have the most questions, since the methods of scientific inquiry and investigation affect everything that is done in science.

ANSWERING SCIENCE TAKS QUESTIONS

There are several different types of questions on the TAKS:

★ Some questions will test your ability to **recall information** by identifying, describing, or giving an example of something. For example:

> **1 Which of the following would be safe during a laboratory activity?**
>
> A Run around the classroom   C Touch a hot surface
> B Leave a spill on the floor   D Obey the teacher's directions

★ Some questions will ask you to **explain**, **compare**, or **identify** a **cause** or **effect**, or place events **in order**.

> **2 Most canyons are formed by —**
>
> F violent tornadoes   H rock sediment
> G river erosion   J volcanic eruptions

★ Some questions will ask you to **read** or **interpret information** from a graph table, or diagram.

★ Some questions will test your ability to **draw a conclusion** or **make a prediction**.

> **3 Salamanders have sticky tongues and wide mouths lined with teeth. These physical characteristics help them to feed on —**
>
> A insects and tiny animals   C algae and small microorganisms
> B leaves and plants   D dead animals

Whatever type of question you are asked, we suggest you follow the same three-step approach to answer it. Think of this as the **E-R-A** approach:

EXAMINE
the question

RECALL
what you know

APPLY
what you know

Let's take a closer look at each of these steps and see how they can help you to choose the correct answer to a TAKS science question:

STEP 1: EXAMINE THE QUESTION

Start by reading the question carefully. Make sure you understand any information the question provides. Look closely at the answer choices. Eliminate any answer choices you know are obviously wrong. Let's use **Question 2** on the previous page to see how this step applies:

> You can see that this question asks you to identify a cause-and-effect relationship: "Most canyons are formed by _____?_____." You should ask yourself: What causes canyons to be created?

STEP 2: RECALL WHAT YOU KNOW

Next you need to identify the subject that the question asks about. Take a moment to think about what you already know about that subject. Mentally review the most important concepts, facts, and relationships that you can remember.

> Now you have to stop and think about what you can recall from your study of Earth Science. The question asks about land forms. Sit back and think hard about what you can recall about this topic. You may remember that a canyon is a narrow valley that cuts through rock. You may even remember that rivers sometimes cut away the land as they flow towards the sea.

STEP 3: APPLY WHAT YOU KNOW

Now take what you can recall about the topic of canyons and land forms and apply it to the question.

> This question asks you to identify what process forms a canyon. Since you might recall that canyons are narrow valleys that have been cut through rock, you need to think — what would cause the rock to be cut away? In thinking about this, you can eliminate **Choice H**: "rock sediment" and **Choice J**: "volcanic eruptions." Both add material to the Earth's surface: they do not cut through it. The choice is between **Choice F**: "violent tornadoes" and **Choice G**: "river erosion." It makes more sense that over time "river erosion" has the power to cut through rock. Therefore, answer **Choice G** is the correct answer.

UNIT 1

THE NATURE OF SCIENCE

What is science? **Science** is a way of investigating and explaining the natural world. The **Elementary Science TAKS** will test your understanding of how scientists work.

The first unit of this book reviews principles common to all fields of science. To understand the world, scientists ask questions, form and test hypotheses, draw conclusions, and share ideas.

★ **Chapter 1: Scientific Investigation**

This chapter focuses on the process of scientific investigation. You will learn how scientific inquiry guides the development of scientific knowledge, how science is based on theories, and how scientists test their theories through investigation. You will also learn about the variety of tools and methods used by scientists when conducting a scientific investigation, and about the importance of safety in experiments.

★ **Chapter 2: Measuring and Analyzing Data**

This chapter looks at how scientists use different instruments to take measurements. You will also learn how scientists analyze data and draw conclusions.

★ **Chapter 3: Critical Thinking in Science**

This chapter looks at how scientists use their observations of the world and data from experiments to develop scientific explanations.

CHAPTER 1

SCIENTIFIC INVESTIGATION

In this chapter, you will learn how scientists think and investigate.

— MAJOR IDEAS —

A. Scientists use different methods and tools to solve problems.

B. The following steps are often used by scientists to conduct experiments:

★ A scientist observes the world and asks a question.
★ The scientist develops a **hypothesis** to answer the question.
★ The scientist designs an experiment to test the hypothesis.
★ The scientist uses special equipment to carry out the experiment.
★ After scientists conduct an experiment, they interpret their results.
★ Scientists communicate their results and conclusions to others.

C. Scientists take safety into account in field and laboratory investigations.

HOW SCIENTISTS WORK

Scientific investigation begins with observation of the natural world. Scientists then ask questions about what they observe. A scientist might watch children fly paper airplanes. The scientist might then ask: How can a paper airplane be made to fly a longer distance? Scientists then try to answer their questions. To answer scientific questions, scientists build models, make new observations and conduct experiments.

THE STEPS OF A SCIENTIFIC INVESTIGATION

Let's look at the steps a scientist might use to answer a scientific question.

ASK A WELL-DEFINED QUESTION

A scientist begins by observing the world. Often what a scientist sees raises one or more questions.

MAKE A TESTABLE HYPOTHESIS

The scientist tries to answer a question with an educated guess, or **hypothesis**. This should be something the scientist can test.

PLAN AN EXPERIMENT

The scientist tests the hypothesis by observing nature or by conducting an experiment.

CHOOSE EQUIPMENT AND TECHNOLOGY

In planning the experiment, the scientist must decide what equipment and technology to use.

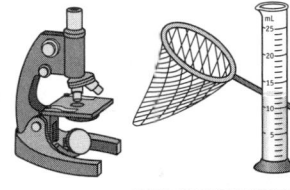

COLLECT INFORMATION

Now the scientist is ready to conduct the experiment. The scientist carefully measures and records the results.

ANALYZE THE RESULTS

The scientist analyzes the information collected from the experiment. Scientists often organize results in a table, graph or chart.

DRAW CONCLUSIONS

The scientist thinks about what the results show. The results should relate to the hypothesis the experiment is testing.

COMMUNICATE THE RESULTS

The scientist communicates the results. The scientist describes the procedures used, so they can be repeated by other scientists.

The order of these steps may sometimes change, and results may cause scientists to form a new hypothesis. Let's look more closely at each of these steps. Suppose you are interested in flying paper airplanes. You want to make your planes fly farther. How would a scientist create an experiment for improving the design of a paper airplane?

ASK A WELL-DEFINED QUESTION

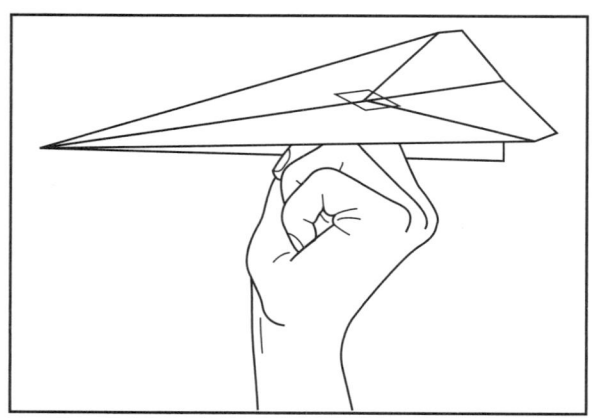

Scientists develop questions based on what they see. Then they define their questions more precisely. Only well-defined questions can be tested by an experiment. Vague questions or ones that ask for opinions cannot be answered by scientific investigation. Questions for investigation must be *specific*, *factual* and identify *exactly* what will be tested in the experiment. For example, the following question is **not** precise enough for a specific experiment: *What is the best paper airplane*?

A scientist would wonder what "best" means. Does it mean the prettiest? Or does it mean the most expensive? A more well-defined question would be:

> *Will a paper airplane fly a longer*
> *distance if it has a flat nose or a pointed nose?*

MAKE A TESTABLE HYPOTHESIS

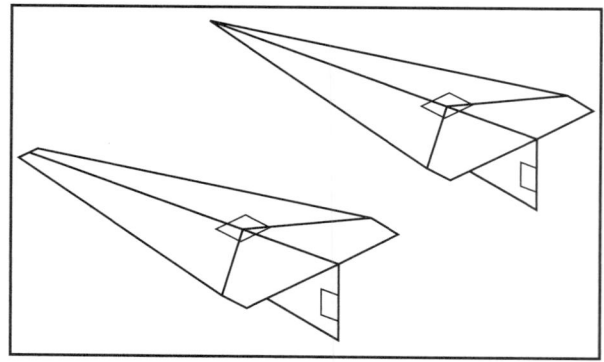

A **hypothesis** is an educated guess that attempts to answer the question. A good hypothesis can be **tested** by an experiment. For example, a scientist may make the hypothesis that a paper airplane will fly farther with a pointy nose than with a flat nose. This prediction can be tested by an experiment.

An experiment may show that the hypothesis is either right or wrong. In science, proving that a hypothesis is wrong can be just as valuable as proving it is right.

APPLYING WHAT YOU HAVE LEARNED

✦ Why is it just as important to prove that a hypothesis is wrong as to prove it is right? _____

_____

✦ Think of an experiment you did in science class this year. What hypothesis did that experiment test? _____

_____

_____

_____

PLAN THE EXPERIMENT

An experiment creates special conditions to test the hypothesis. It is important to plan the experiment carefully.

★ **Variables.** A variable is anything that can change in the experiment. In most experiments, a scientist changes one thing or variable to see what effect this has on something else. The scientist then observes or measures the effect of this change. For example, a researcher conducting an experiment on paper airplanes can change these variables:

Experiments deal with independent and dependent variables.

> **What kind of paper should the airplane be made of?**

> **What is the shape of the paper airplane?**

★ In an experiment, a scientist usually changes only **one variable** at a time. For example, in the experiment involving paper airplanes, a scientist will pick one variable — such as whether the plane has a flat or pointy nose. The scientist will change **only** this variable and see how that affects something else — such as how far the airplane will fly. All other conditions will be kept exactly the same. Each plane will be the same size, be made of the same type of paper, and have the same shape except for the nose.

How do changes in one variable affect a second variable?

Change in Variable A → Impact of this change on Variable B

Sometimes scientists have two groups in an experiment. The scientist changes something for one group (**experimental group**) but not for the other group (**control group**). Then the scientist compares the results of the two groups.

APPLYING WHAT YOU HAVE LEARNED

Examine the following information:

Experimental Group	Students are given vitamin C when they have a cold.
Control Group	Students are not given vitamin C when they have a cold.

Members of both groups record the number of days each cold lasts.

★ What is being investigated?_____

★ Why was a control group necessary in this experiment?_____

_____

APPLYING WHAT YOU HAVE LEARNED

Look at the following examples. For each example, identify the question that the scientist is investigating:

Variable the Scientist Changes	Variable the Scientist Measures	What question is the scientist trying to answer?
Amount of sugar someone eats	The time the person goes to bed	
Type of nose a paper airplane has	Distance the paper airplane can fly	
Amount of oxygen in a container	Amount of time a candle in the container will burn	
The direction of grooves on a hill	Amount of soil erosion in a heavy rainfall	

USE EQUIPMENT AND TECHNOLOGY

Planning an experiment is like making a recipe. First you must identify the materials, equipment and technology that are needed. Then you must list the steps to be followed to conduct the experiment.

ELEMENTS OF A GOOD EXPERIMENTAL DESIGN

★ The experiment tests the hypothesis.

★ All the variables are identified.

★ All required materials and equipment are listed.

★ Results can be precisely measured.

★ There should be several trials.

A. To begin this experiment, you need: (1) two pieces of paper identical in size; (2) a roll of transparent tape; (3) a meter stick. Start by making two paper airplanes with pointy noses that are exactly the same. Here is how it is done:

Step 1:	**Step 2:**	**Step 3:**	**Step 4:**

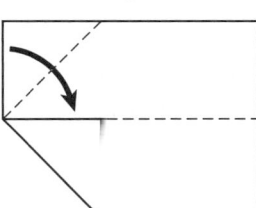 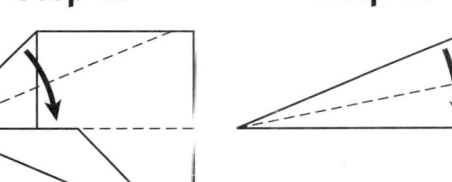 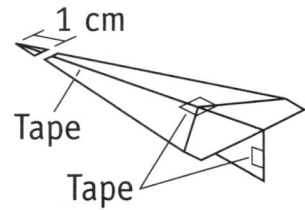

B. Then cut off the first 1 cm. piece from one of your two planes, giving it a "flat" nose. Attach the piece of paper you cut off into the fold in the center of that plane, so that its weight stays the same.

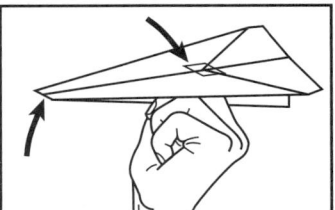

C. Now you are ready to test your airplanes. Toss each plane from an identical location using the same force.

D. Measure and record the distance each paper airplane travels after it lands.

E. Repeat the experiment several times with each paper airplane.

Type of Plane	Distance: Trial #1	Distance: Trial #2	Distance: Trial #3	Distance: Trial #4
"Pointy-Nosed Plane"				
"Flat-Nosed Plane"				

STANDARD LABORATORY AND FIELD EQUIPMENT

Experiments often require equipment. The following are some of the laboratory and field equipment you should know:

Equipment for Taking Measurements

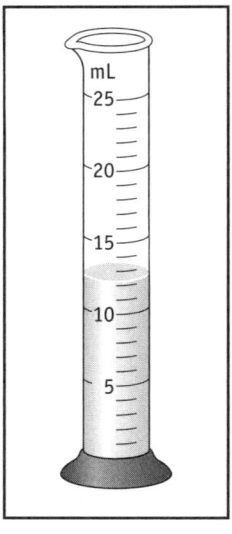

★ **Meter Stick.** A ruler marked in centimeters (cm.), used to measure length.

★ **Graduated Cylinder.** A glass cylinder marked in milliliters (mL), used to measure the volume of liquids.

★ **Thermometer.** An instrument used to measure temperatures in degrees.

★ **Balance.** An instrument with one or two pans, used to measure the mass of an item in grams (g) or kilograms (kg).

★ **Timing Devices.** Clocks, stop watches or other devices that precisely measure the passage of time in seconds, minutes, and hours.

Equipment for Safety

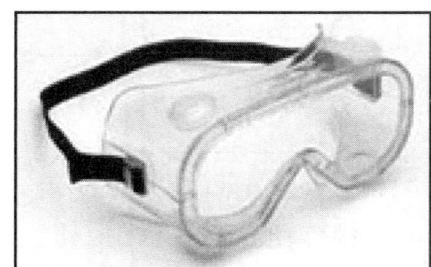

★ **Safety Goggles.** Plastic goggles large enough to protect the eyes and face during an experiment from fine dust, splashes, mists, or sprays.

★ **Laboratory Aprons.** Bibs worn over clothing to protect clothing and the skin from splashes or spilled chemicals or biological materials.

Other Laboratory Equipment

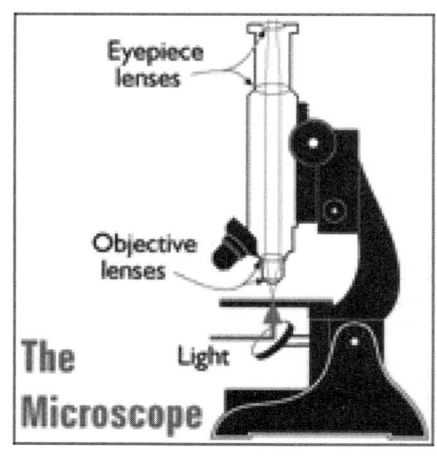

★ **Microscope.** An instrument that uses a series of lenses to magnify specimens placed on slides.

★ **Hand Lens.** A magnifying glass used to inspect the features of something more closely.

★ **Magnet.** A piece of iron that attracts other iron objects and also affects electrical currents. Magnets are used for experiments on magnetism and electricity.

Eyepiece lenses

Objective lenses

The Microscope

Light

Field Equipment

★ **Collecting Nets.** Nets used to collect field samples.

★ **Rain Gauges.** Collects and measures the amount of water when it rains.

APPLYING WHAT YOU HAVE LEARNED

Equipment	How It Looks	What Is It Used For?
1. Safety goggles		
2. Microscope		
3. Hand lens		
4. Collecting net		
5. Balance		
6. Thermometer		
7. Stopwatch		
8. Rain gauge		
9. Graduated cylinder		
10. Magnet		

SAFETY PRECAUTIONS

Attention to safety is essential during both field and laboratory investigations. Safety must be considered even before the experiment begins.

APPLYING WHAT YOU HAVE LEARNED

Look at the following common laboratory safety rules. Examine the list in the first column and then explain why each rule is important for safety.

Laboratory Safety Rules	Why this Rule is Important for Safety
Read all procedures before starting a laboratory investigation. Know what hazards you may face.	
Don't begin an experiment until your teacher has given directions.	
Wear safety equipment when working with chemicals or liquids that could spill or splash.	
Follow directions at all times. Don't smell or taste anything unless directed by your teacher.	
Follow all steps, procedures or directions exactly when you conduct an experiment.	
If you are heating liquids, point them away from yourself or other students.	
If an accident occurs, tell your teacher about it immediately.	
Clean up your work area and put materials away after you are done.	
Wash your hands with soap and water before and after all experiments.	

APPLYING WHAT YOU HAVE LEARNED

How good are you at identifying some of the basic safety signs often found in a classroom laboratory? Identify each of the safety signs below:

_____   _____   _____   _____   _____

COLLECT DATA

During an experiment, scientists often take precise measurements with special equipment. Information collected during an investigation is known as **data**.

ANALYZE THE DATA

Once data has been gathered from an experiment, the scientist studies it to see what it shows. Scientists often make tables, charts, and graphs to help them see patterns in the data. You will learn how scientists analyze data in the next chapter.

APPLYING WHAT YOU HAVE LEARNED

Suppose you wanted to design an experiment to see the effect of ABC's fertilizer on the growth of lima bean plants. The materials for your experiment are two large flower pots with soil, lima bean seeds, a sunny garden, water, a watering can, a meter stick, and a bag of ABC's "Miracle Fertilizer."

◆ What variable would you change? _____

◆ What steps would you take to carry out your experiment?

  • _____

  • _____

  • _____

◆ How would you measure the results of your experiment? _____

  _____

FORM CONCLUSIONS

Scientists draw **conclusions** from their analysis of the data. Their conclusions must be supported by scientific knowledge and evidence from the investigation. The conclusion should support or reject the hypothesis, or suggest changes in the hypothesis for further study.

APPLYING WHAT YOU HAVE LEARNED

◆ Why do you think it is important that the conclusions of an experiment should relate back to the hypothesis? _____

_____

_____

COMMUNICATE RESULTS

Once a scientist completes an experiment, the results must be communicated to others. Usually this takes place in the form of a written report, article, or an oral presentation. A good presentation uses clear language, includes accurate data and graphs, and includes the exact procedures followed by the scientist so other scientists can *repeat* the experiment and check the results.

A good scientist communicates results to other scientists orally or in a written report.

APPLYING WHAT YOU HAVE LEARNED

◆ Why is it so important that the procedures of an experiment reported by one scientist can be repeated by other scientists? _____

_____

_____

WHAT YOU SHOULD KNOW

A. You should know that science asks questions about the natural world and tries to answer them by using special scientific methods.

B. You should know that scientists use certain steps to carry out an experiment:

◆ The scientist observes nature and asks well-defined questions.

◆ From these observations, the scientist will develop a **hypothesis** or educated guess to try to answer the question.

◆ The scientist will then design an experiment to test the **hypothesis**. Often the experiment changes one **variable** to see what effect this has on another **variable**. Other variables are held constant.

◆ Scientists measure their results. Laboratory experiments help them take exact measurements.

◆ Scientists analyze their data and draw conclusions. Their conclusions should relate to the original hypothesis.

◆ Scientists communicate their results to others.

C. In conducting an experiment or investigation, you should always take safety into account.

CHAPTER STUDY CARDS

Steps in a Scientific Investigation

★ **Ask a well-defined scientific question**

★ **Form a testable hypothesis**

★ **Design an experiment to test the hypothesis**

★ **Select and use equipment and technology**

★ **Collect data**

★ **Analyze data**

★ **Form conclusions**

★ **Communicate results**

Laboratory and Field Terminology

★ **Hypothesis.** An educated guess that tries to answer a question under investigation.

★ **Variable.** Something that can be changed or varied to find how that change affects other things in the experiment.

★ **Laboratory Equipment.** These tools include a balance, meter stick, thermometer, hand lens, graduated cylinder.

★ **Laboratory Safety.** Safety should always be an important concern in all experiments.

APPLYING WHAT YOU HAVE LEARNED

◆   Create your own chart showing the various steps of a scientific investigation. Make a separate box for each step. Describe each step and draw a picture to illustrate it.

CHECKING YOUR UNDERSTANDING

1   **Juan gives one group of 10 chickens a type of chicken feed that has little protein. He gives a second group of 10 chickens the same amount of chicken feed each day, but adds a small amount of protein powder to the chicken feed. He weighs each group of chickens at the beginning of the experiment. Two months later he weighs the chickens a second time.**

What question is Juan trying to answer in this experiment?

A   Will protein powder cause chickens to live longer?

B   Does protein powder cause chickens to gain weight?

C   Do chickens prefer chicken feed with protein powder?

D   Can chickens become stronger by eating protein powder?

OBJ. 1
5.2 (A)

HINT

This question looks at the methods used by scientists to carry out a scientific investigation. You should understand that an experiment usually looks at the effects that changing one thing or variable has on another variable. In this question, only one variable has been changed — chickens have had protein powder added to their feed. The scientist then weighs the chicken. The experiment tests whether eating protein powder will cause chickens to gain weight. Therefore, the answer is **B**.

Now try answering some additional questions on your own:

2 **Keisha wants to measure the time it takes her to run fifty meters. What instrument should Keisha use?**

F   A meter stick

G   A graduated cylinder

H   A collecting net

J   A stop watch

OBJ. 1
5.2 (A)

3 **Sarah Jones adds vinegar to a container of baking soda in her science class. Bubbles of carbon dioxide form. What safety precaution should Sarah take while conducting this experiment?**

A   List the materials at the end of the experiment

B   Wear safety goggles and gloves

C   Wash her eyes with eyewash

D   Leave the results of the experiment for the next class to clean up

OBJ. 1
5.1 (A)

> *A student places a wooden spoon and a stainless steel spoon in a container of boiling water. After placing the spoons in the boiling water, she waits five minutes and then measures the temperature at the end of the handle of each spoon.*

4 **Which instrument should the student use to measure the temperature of the spoon handles?**

F   A thermometer

G   A balance scale

H   A meter stick

J   A rain gauge

♦ **Examine the Question**
♦ **Recall What You Know**
♦ **Apply What You Know**

OBJ. 1
5.2 (A)

5 **In the experiment described above, what question do you think the student is trying to answer?**

A   Is steel a better conductor of heat than wood?

B   Does wood conduct electricity better than steel?

C   Can water conduct heat?

D   Does heat cause either wood or steel to expand?

OBJ. 1
5.2 (A)

6 **Wendy wants to find out how much rain will occur during the next storm. Which tool should Wendy use to collect her data?**

F   A rain gauge

G   A thermometer

H   A balance scale

J   A stop watch

OBJ. 1
5.2 (A)

7   James has decided to investigate whether the number of flowers on a plant will increase if water supplied to that plant is increased. James has four pots of geraniums to use during his experiment.

What factor in the experiment should be varied for the four plants in order to answer the question?

A   Temperature of the water   C   Age of the seeds

B   Number of hours in sunlight   D   Amount of water

OBJ. 1
5.2 (A)

8   Alice places a magnet next to a metal fork. She records what happens. Next, she places the same magnet next to a fork made of plastic. She records what happens next.

Which question is Alice most likely exploring with this experiment?

F   Does the size of the magnet affect the magnet's power?

OBJ. 1
5.2 (A)

G   What causes magnets to be attracted to metal objects?

H   How do different objects react to magnets?

J   Do objects other than magnets have magnetic power?

9   Kara did an experiment to find out the effect that temperature has on the activity of yeast. Which step would come last in Kara's experiment?

A   Move one bowl to a warmer location.

OBJ. 1
5.2 (A)

B   Observe what happens in each bowl.

C   Put 120 milliliters of water in each bowl.

D   Add 1 gram of yeast to each bowl.

10 These fifth-graders are doing a science experiment at their work station during science class. Which of these students is NOT practicing good laboratory safety?

OBJ. 1
5.1 (A)

F   G   H   J

CHAPTER 2

MEASURING AND ANALYZING DATA

In this chapter, you will learn how scientists take precise measurements and analyze the data they collect.

— MAJOR IDEAS —

A. Scientists collect information by observing and measuring.

B. Scientists use meter sticks, timing devices, balances, graduated cylinders, and thermometers to take precise measurements.

C. Scientists make simple graphs, tables, maps and charts to help them organize, examine and evaluate the information they collect.

D. Scientists interpret data to draw conclusions.

In the last chapter, you learned that scientists collect data during experiments and field investigations. Sometimes scientists just make simple **observations** of what they see, hear, feel, taste or smell. Very often, scientists **measure** the results of an experiment.

UNITS OF MEASUREMENT

Scientists use the **metric system** to make measurements.

milli = one thousandth	**centi** = one hundredth	**kilo** = one thousand

★ **Length.** A meter is just over 3 feet long.
  • 10 millimeters (mm) = 1 **centimeter** (cm) • 1,000 meters (m) = 1 **kilometer** (km)
  • 100 centimeters (cm) = 1 **meter** (m)

★ **Mass.** A kilogram is about 2.2 pounds.
  • 1,000 grams (g) = 1 **kilogram** (kg)

★ **Volume.** A liter is just a little more than 1 quart.
  • 1,000 milliliters (mL) = 1 **liter** (L)
  • 1 cubic centimeter (cm^3) = 1 **milliliter** (mL)

★ **Temperature.** Scientists measure temperature in degrees Celsius.
  • 100° C = one hundred degrees Celsius (*temperature at which water boils*)
  • 0° C = zero degrees Celsius (*temperature at which water freezes*)

MEASURING LENGTH

You will be given a 20 cm paper ruler on the TAKS test similar to the ruler on the margin of the next page. To find the length of something, find the 0 mark of the ruler. Then put your object starting at this mark. Look at where the object ends on your ruler. Write down the number that

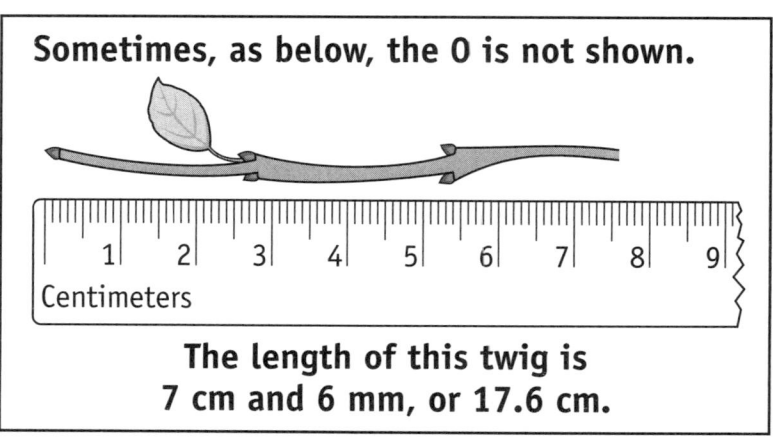

Sometimes, as below, the 0 is not shown.

The length of this twig is 7 cm and 6 mm, or 17.6 cm.

is just below that ending point. Then count the narrow lines between the zero and the end of your object. This tells you how many millimeters to add. You can add the centimeters and millimeters together by using a decimal point: **centimeters. millimeters**. This gives you the length in centimeters (cm).

APPLYING WHAT YOU HAVE LEARNED

What is the length of this worm?

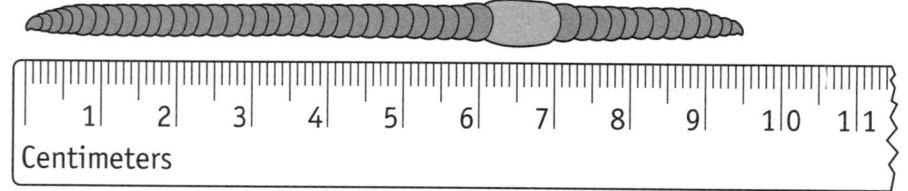

_____ cm + _____ mm. Add these two numbers together.

How many centimeters is the earthworm? _____ cm

BUBBLING IN YOUR ANSWERS

Some questions on the **TAKS Grade 5 in Science** may require you to write a number instead of picking an answer choice. After you have examined the question and found what you believe to be the correct answer, you will need to enter that answer number on a grid or bubble format just like the ones below.

All of the answers will be whole numbers, with the last column showing the placement of the decimal point. The left column is for hundreds, the next column is for tens, and the right column is for single digits (units). You must record your answer in the columns for the correct place values. The second grid on the right shows you how the number 47 would look when it is bubbled in. Notice that the hundreds column has a "0."

			.
⓪	⓪	⓪	
①	①	①	
②	②	②	
③	③	③	
④	④	④	
⑤	⑤	⑤	
⑥	⑥	⑥	
⑦	⑦	⑦	
⑧	⑧	⑧	
⑨	⑨	⑨	

0	4	7	.
●	⓪	⓪	
①	①	①	
②	②	②	
③	③	③	
④	●	④	
⑤	⑤	⑤	
⑥	⑥	⑥	
⑦	⑦	●	
⑧	⑧	⑧	
⑨	⑨	⑨	

APPLYING WHAT YOU HAVE LEARNED

Now use your own centimeter ruler or cut out the 20 cm paper ruler on this page. Then measure the objects below and complete the grids:

A. What is the length of the paper clip?

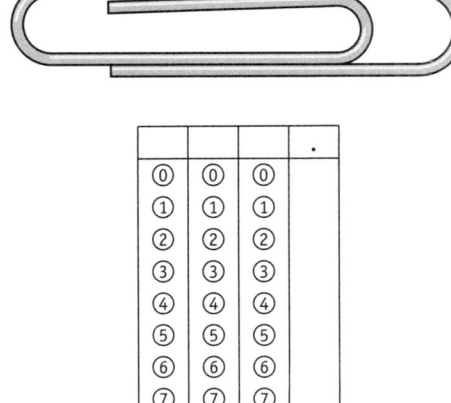

			.
⓪	⓪	⓪	
①	①	①	
②	②	②	
③	③	③	
④	④	④	
⑤	⑤	⑤	
⑥	⑥	⑥	
⑦	⑦	⑦	
⑧	⑧	⑧	
⑨	⑨	⑨	

B. What is the length of this leaf from point A to point B

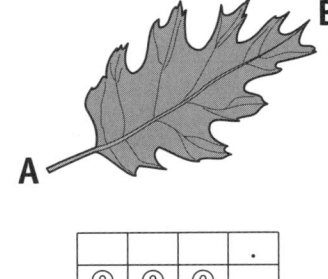

			.
⓪	⓪	⓪	
①	①	①	
②	②	②	
③	③	③	
④	④	④	
⑤	⑤	⑤	
⑥	⑥	⑥	
⑦	⑦	⑦	
⑧	⑧	⑧	
⑨	⑨	⑨	

Centimeters 0 1 2 3 4 5 6 7 8 9 10 11 12 13 14 15 16 17 18 19 20

MEASURING VOLUME

Volume measures how much space something takes up. To measure the volume of a liquid, scientists pour the liquid into a graduated cylinder. The cylinder usually has lines for each milliliter (mL) up to 100 milliliters (mL). The surface of the water curves up the sides of the cylinder. Measure the volume of the liquid from the flat bottom of the curve. See what line is the closest on the side of the cylinder.

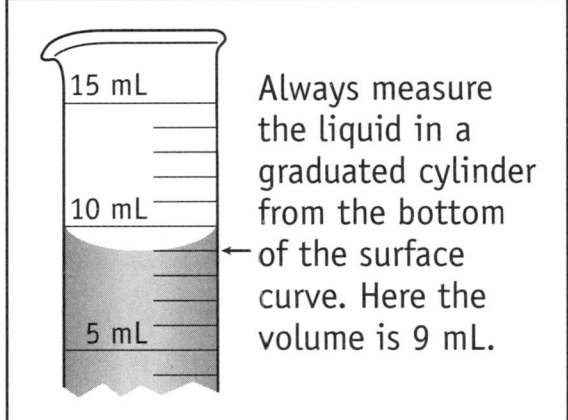

Always measure the liquid in a graduated cylinder from the bottom of the surface curve. Here the volume is 9 mL.

To measure the volume of a small solid, like a rock, add water to a graduated cylinder. Record the volume of the water. Now put the solid into the graduated cylinder. Record the new volume. Subtract the original volume of water from the new volume. The volume of a solid object is usually recorded in cubic centimeters. Each $1mL = 1cm^3$

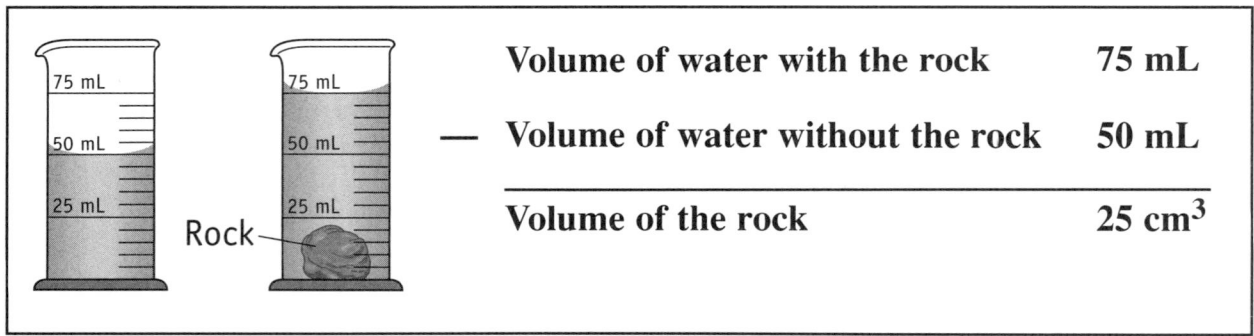

Volume of water with the rock	75 mL
— Volume of water without the rock	50 mL
Volume of the rock	25 cm^3

APPLYING WHAT YOU HAVE LEARNED

Two graduated cylinders each have the same amount of water. What is the volume of the stone in the second graduated cylinder?

_____ cm^3

APPLYING WHAT YOU HAVE LEARNED

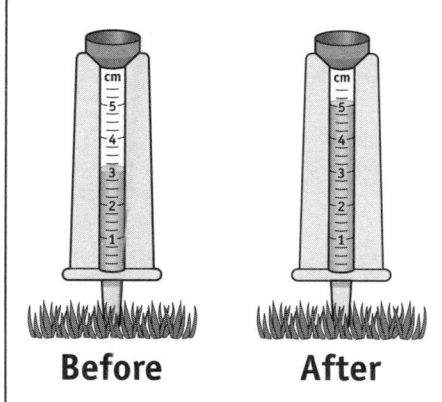

Before   **After**

A **rain gauge** looks like a graduated cylinder. However, instead of measuring volume, it measures how much rain has fallen as a length in centimeters. The picture to the left shows the same rain gauges before and after a rainstorm.

How many centimeters of rain fell during the rainstorm? _____

MEASURING MASS

Mass is the amount of matter an object has. The weight of an object is closely related to its mass. Scientists use a **balance** to measure mass. A balance puts a known mass on one side and the object to be measured on the other side.

A DOUBLE-PAN BALANCE

A **double-pan balance** has a bar with pans on each side. The scientist puts the object that is to be measured in one pan. In the other pan, the scientist puts known units of mass. Units of mass are added until the bar is level — both pans are the same height. The scientist then adds up all the known units to find the mass of the object.

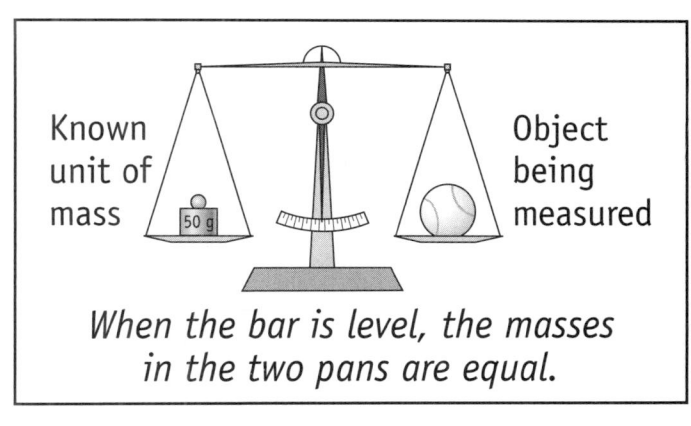

When the bar is level, the masses in the two pans are equal.

APPLYING WHAT YOU HAVE LEARNED

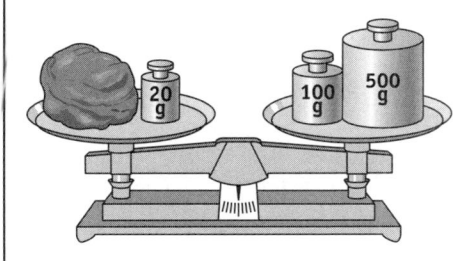

The pans in this double-pan balance are at the same height. What is the mass of the rock being measured on the left side?

_____ grams

A TRIPLE-BEAM BALANCE

A **triple-beam balance** has a single pan. Three scaled beams with "riders" are used to measure the mass of whatever is placed in the pan. The riders are moved along notches on each beam.

To find the mass of the object in the pan, add up the numbers shown on the three riders. For example, a sample is placed on a triple-beam balance. The picture below shows the riders when the scale is balanced. What is the mass of the sample being measured? _____

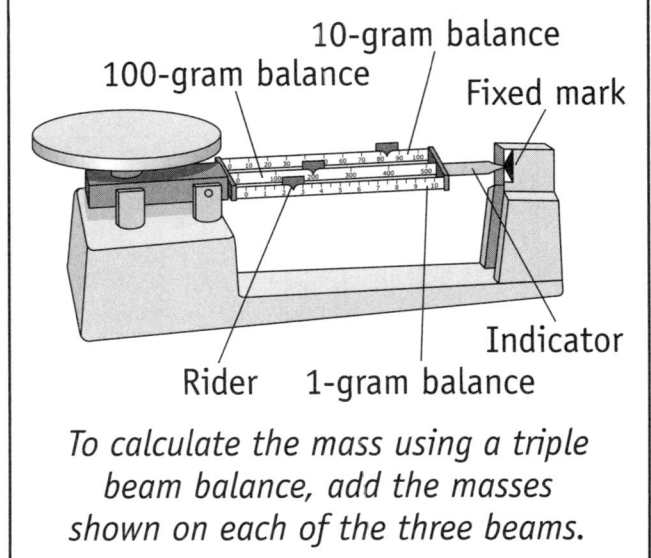

To calculate the mass using a triple beam balance, add the masses shown on each of the three beams.

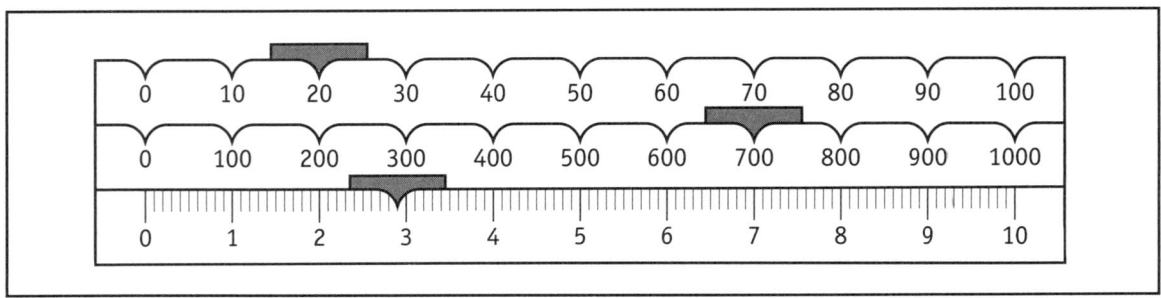

APPLYING WHAT YOU HAVE LEARNED

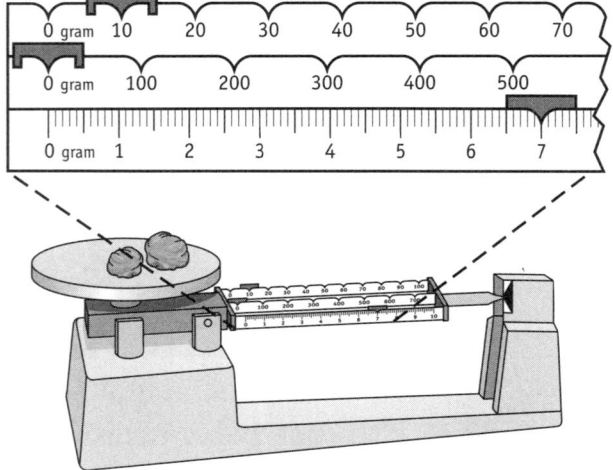

What is the mass of the rocks on this triple-beam balance? _____

MEASURING TEMPERATURE

Temperature is a measure of how hot or cold an object is. To measure temperature, scientists use a **thermometer**. There are many kinds of thermometers. Many thermometers are glass tubes with colored liquid inside and numbers printed along the tube. As the thermometer gets warmer, the liquid expands and rises inside the tube. To measure the temperature, look at the *highest point* the liquid has reached. Be sure to keep your eye at the same level as the liquid. Then read the number of degrees. Scientists use the Celsius scale to measure temperature. Water freezes at 0° C and boils at 100° C.

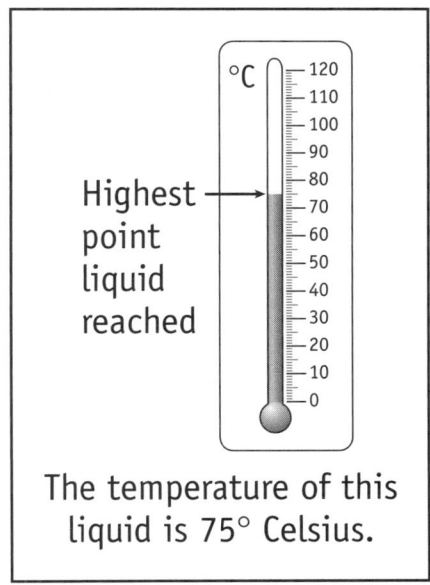

The temperature of this liquid is 75° Celsius.

APPLYING WHAT YOU HAVE LEARNED

What is the temperature of each of the following thermometers?

1. _____ °C

2. _____ °C

3. _____ °C

Which of these shows the boiling point of water? _____

ANALYZING DATA

Once scientists take measurements, they record their results. Then they look for patterns in the data. To help find patterns, scientists often arrange their results in the form of a table, graph, chart or map. You should know how to make and interpret each of these different forms of data.

TABLES

A **table** is used to organize information. Tables list information in columns and rows. To interpret a table, you need to pay close attention to its headings. A scientific table often shows the relationship between two variables that are being measured.

For example, look at the table below. It is based on the experiment with paper airplanes discussed in the last chapter. This table shows the relationship between the type of paper airplane and how far it flew:

| TYPE OF AIRPLANE | DISTANCE AIRPLANE FLEW | | | |
	Trial 1	Trial 2	Trial 3	Trial 4
"Flat-nosed" Airplane	4 m	5 m	4 m	3 m
"Pointy-nosed" Airplane	6 m	5 m	7 m	6 m

APPLYING WHAT YOU HAVE LEARNED

◆ What distance did the flat-nosed airplane fly on its third trial? _____

◆ What was the average distance (*add the numbers, then divide the total by the amount of numbers*) flown by the pointy-nosed airplane on each trial?
_____

◆ What conclusion can you draw from this data? _____
_____
_____
_____

BAR GRAPHS

A **bar graph** is made up of bars of different lengths. It is used to compare data. Each bar represents a quantity of something. Each bar is labeled or a key is provided to tell what each bar represents. Look at the bar graph on the following page. It shows the average distance flown by each of the two types of paper airplanes — the pointy-nosed and flat-nosed airplane.

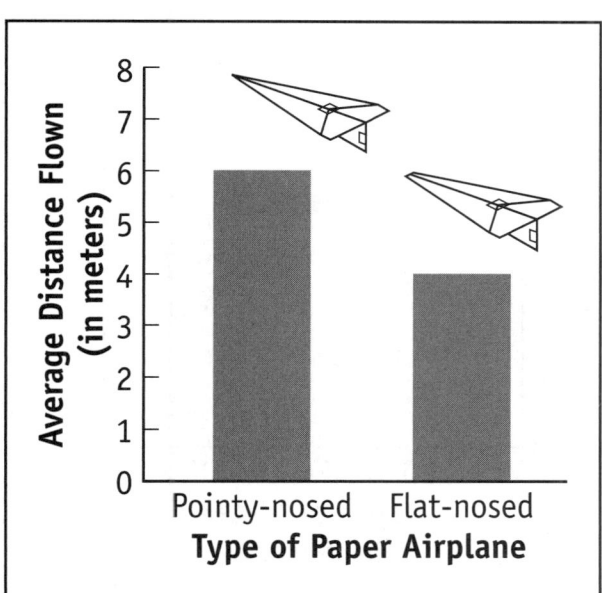

★ What is the average distance flown by the flat-nosed airplane? _____

★ Which of the airplanes was able to fly a greater distance? _____

★ How is a bar graph different from a table? _____

_____

_____

_____

LINE GRAPHS

A **line graph** shows a series of connected points on graph paper or a similar grid. Each point on the grid represents a quantity. A line graph is usually labeled along its bottom line and left side. Examine the following line graph:

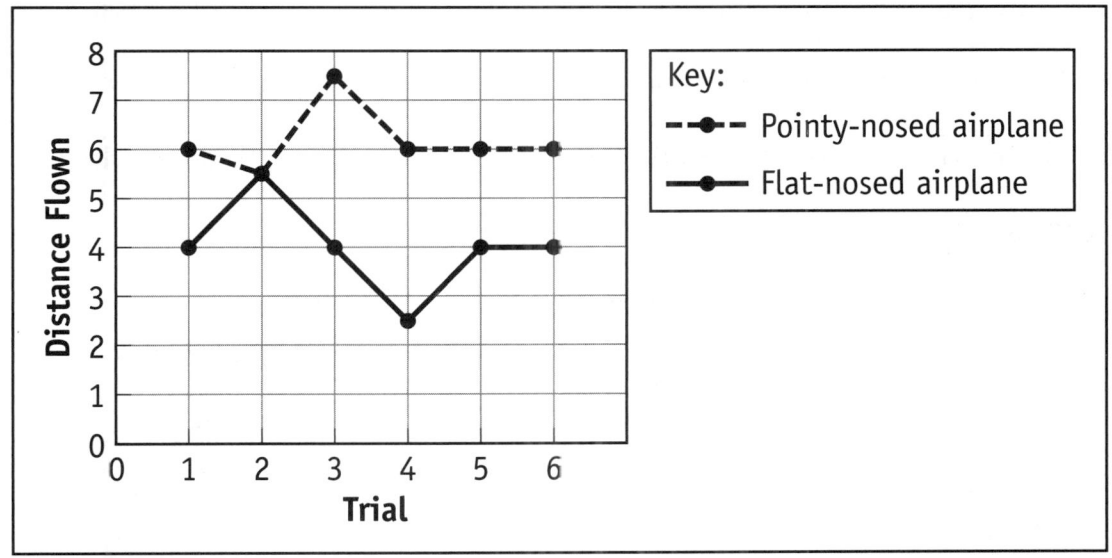

★ Based on the graph, how far did the pointy-nose airplane fly on trial 6? _____

★ Why do scientists often use several trials when they conduct experiments?

_____

_____

_____

Often a line graph is used to show how one variable changes when another variable is altered. Examine the line graph on the right. It shows how the temperature of the ocean changes as you go deeper into the water.

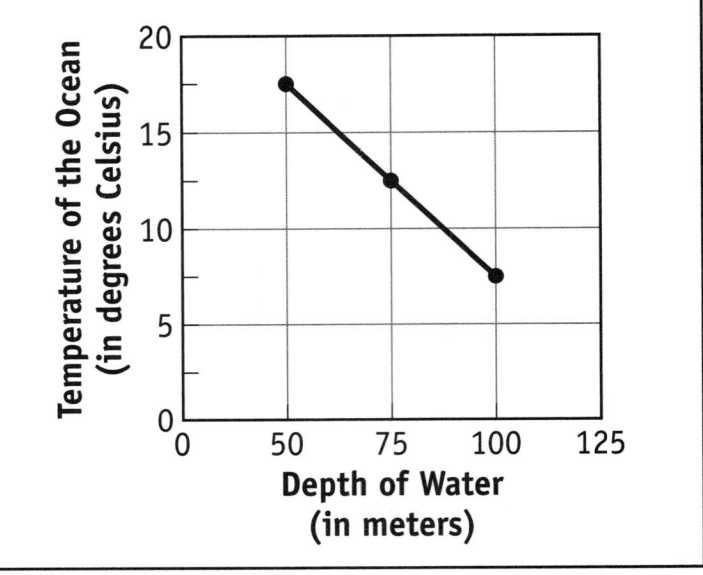

★ What is the temperature of the water at 75 meters deep?

_____

★ What conclusion can you draw from the data presented in this line graph?

_____

_____

_____

Sometimes you can use a line graph to guess at missing variables. For example:

★ What do you think the temperature of the ocean is at 60 meters deep? _____

★ What do you think the temperature of the ocean is at 90 meters deep? _____

PIE CHARTS

Pie charts, also known as **circle graphs**, show how different parts of something relate to the whole. They are also used to show percentages. When all the "pieces of the pie" are added together, they will equal 100%.

The pie chart to the right shows Jack's baseball team. The team has 3 boys and 6 girls. Complete the chart by shading in three "slices."

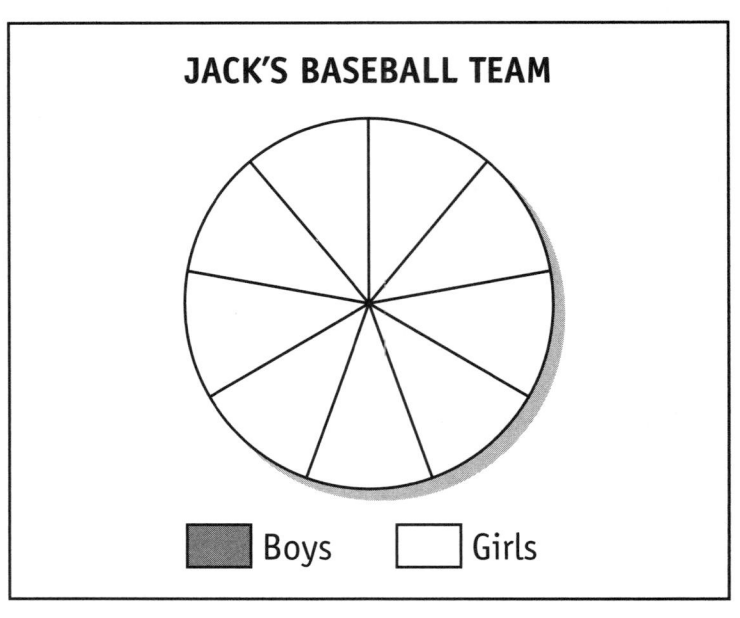

MAPS

A **map** is a diagram representing a place. It shows where objects are located. Maps may be used to show the location of the stars, or to show features on the Earth's surface. The **legend** of the map explains symbols used on the map. The **direction indicator** or **compass rose** shows directions (N, E, S, W) on the map. Now let's look at a map showing part of our solar system. Notice that there is no direction indicator. Examine the map and then answer the questions that follow.

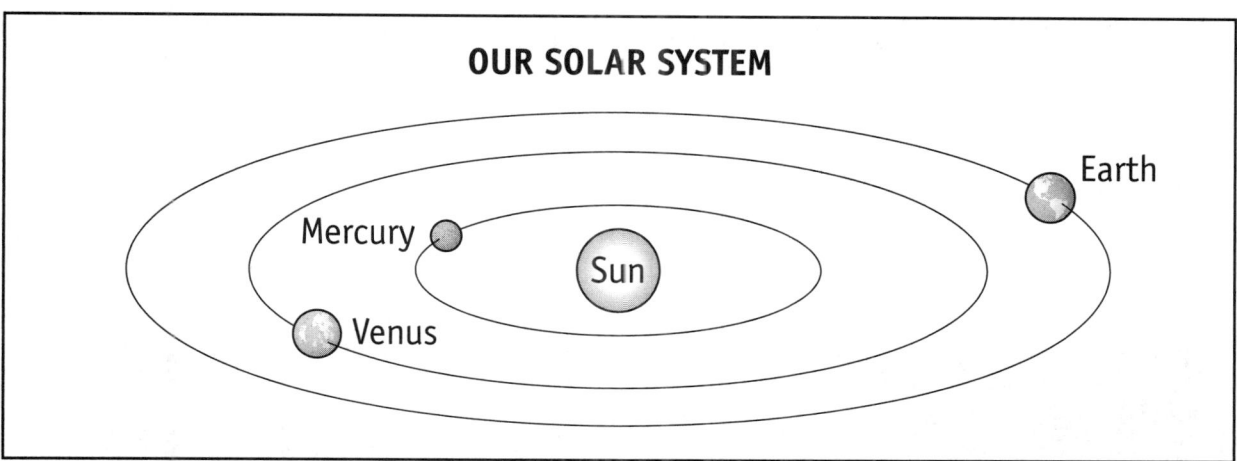

OUR SOLAR SYSTEM

APPLYING WHAT YOU HAVE LEARNED

◆ Which of the three planets on the map is farthest from the sun? _____

◆ Which of the three planets on the map is closest to the sun? _____

◆ Which of the three planets probably has the hottest surface in daytime? _____

◆ Based on the map, which of the three planets most likely takes the longest time to orbit the sun? _____ Explain your answer. _____

_____

_____

_____

CHARTS AND DIAGRAMS

Charts and diagrams can take a variety of different forms. Most use pictures to show how things relate to one another. Lines or arrows often indicate relationships. Look at the diagram presented on the following page.

PRAIRIE FOOD CHAIN

Wheat plants → Mice → Snakes → Owls

This diagram shows that mice eat wheat plants for food. Snakes then eat the mice. The diagram shows the relationship of these plants and animals on the prairie.

APPLYING WHAT YOU HAVE LEARNED

◆ What prairie animal eats snakes for food? _____

◆ Based on the food chain, what do snakes eat for food on the prairie? _____

DRAWING CONCLUSIONS

Many questions on the **Grade 5 TAKS in Science** will ask you to locate specific information on a graph, table or diagram. Other questions will ask you to go beyond finding information. These questions will ask you to draw a conclusion. "Conclusion" questions are usually of two kinds:

| Make a generalization from the data | Make a prediction from the data |

MAKING A GENERALIZATION

A **generalization** is a general statement that summarizes what several specific pieces of data show. For example, look at the graph to the right. This is the same line graph you examined earlier in the chapter.

From this graph, you can tell that at 75 meters deep, the ocean temperature is 13°C. This is specific information.

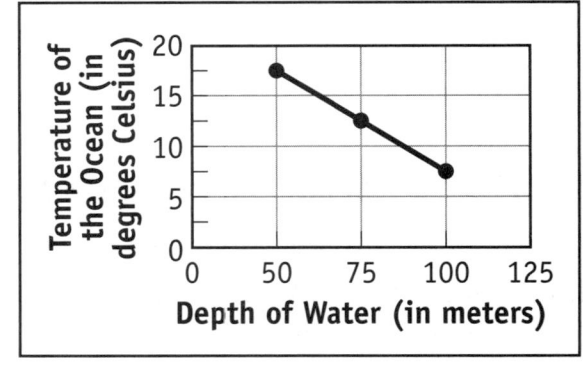

Look again at the temperatures shown by the graph. Can you make a **generalization** about them — a statement or conclusion that applies to a number of examples? You might notice that as you go deeper into the ocean, the water temperature drops or gets colder. This general statement describes what several specific pieces of data on the graph, taken together, show.

MAKING A PREDICTION

It is also possible to make predictions from the data. A **prediction** is a statement of what someone thinks will happen in the future. For example, we can see that the deeper you go into the ocean the colder the temperature becomes. Based on this generalization, we can predict that the temperature of the ocean will be lower at 125 meters than at 100 meters. We can guess that the temperature is likely to be around 3°C.

Based on the information in the graph, what do you think the temperature of the ocean is at a depth of 65 meters? This temperature should be **less** than the temperature at 50 meters, but **greater** than the temperature at 75 meters.

APPLYING WHAT YOU HAVE LEARNED

A scientist conducted an experiment growing nine bean plants in different amounts of light. The results of the experiment are shown below:

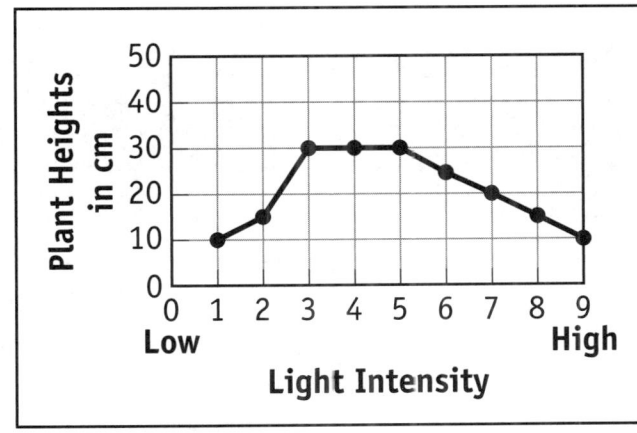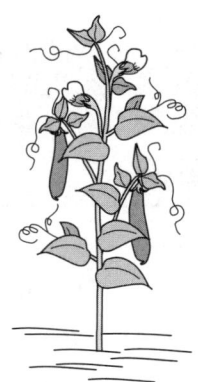

Based on this data, what conclusion should the scientist draw from this experiment?

_____

_____

_____

WHAT YOU SHOULD KNOW

A. You should know that scientists use specific tools to take measurements.

  ✦ To measure **length**, use a meter stick or centimeter ruler. You should know how to use a 20 cm ruler on the TAKS test.

  ✦ To measure **volume**, use a graduated cylinder.

  ✦ To measure **mass**, use a double-pan or triple-beam balance.

  ✦ To measure **temperature**, use a thermometer.

B. You should know that scientists analyze data by making tables, bar graphs, line graphs, pie charts, maps and diagrams. You should be able to read and interpret each of these:

  ✦ **Graphs** often show the relationship between two variables.

  ✦ **Maps** show where things are located.

  ✦ **Charts** show the relationship of parts to a whole.

C. You should know how to draw conclusions from data.

  ✦ To make a **generalization**, look at the data and decide what it shows.

  ✦ To make a **prediction**, guess what the data would show based on what you already know.

CHAPTER STUDY CARDS

Measuring Data

★ **Metric System**
  • **Length:** mm, cm, m, km.
  • **Volume:** mL, cm^3
  • **Mass:** g, kg
  • **Temperature:** degrees Celsius (°C)

★ **Tools**
  • **Length:** meter ruler
  • **Volume:** graduated cylinder
  • **Mass:** double pan or triple-beam balance
  • **Temperature:** thermometer
  • **Rainfall:** rain gauge

Analyzing Data

★ **Ways of Displaying Data.**
  • **Table**   • **Pie Chart**
  • **Bar Graph**   • **Map**
  • **Line Graph**   • **Diagram**

★ **Drawing Conclusions.**
  • **Generalization.** Describe what the data shows by looking at several examples in the data.
  • **Prediction.** Guess what the data would show based on what you already know and learn from the data.

CHECKING YOUR UNDERSTANDING

1 From the chart shown on the right, scientists predict that over the next ten years the population of Brazil will most likely —

A show a decrease

B remain the same

C triple in size

D continue to increase steadily

OBJ. 1
5.3 (A)

POPULATION OF BRAZIL 1970–1995

Number of People (in millions) vs. *Year*

HINT

This question looks at the methods used by scientists to draw conclusions. The data shows a steady increase in the population of Brazil from 1970 to 1995. Based on this trend, scientists are most likely to predict that the population will continue to increase steadily. Therefore, the best answer is **D**.

Now try answering some additional questions on your own:

2 Five milliliters of vinegar are added to water in a graduated cylinder. What will be the total volume of the liquid?

F 27 mL

G 30 mL

H 35 mL

J 37 mL

OBJ. 1
5.2 (B)

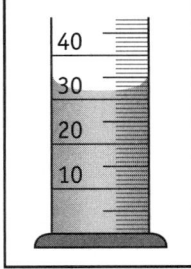

3 Which race is the same length as a 1,000-meter race?

A A 1,000-kilometer race

B A 1,00-kilometer race

C A 10-kilometer race

D A 1-kilometer race

◆ Examine the Question
◆ Recall What You Know
◆ Apply What You Know

OBJ. 1
5.2 (B)

THE GROWTH OF BEAN SEEDS

Temperature (°C)	Days for Seeds to Germinate
25	5
20	7
15	9
10	11
5	?

4 The chart above shows the time it took for bean seeds to germinate at different temperatures. Based on this data, seeds at 5°C will probably germinate in —

F 5 days   **H** 13 days

G 8 days   **J** 16 days

OBJ. 1
5.2 (C)

5 Which metric measurement is closest to the height of the plant to the right?

A 2.5 centimeters

B 5 centimeters

C 7.5 centimeters

D 10 centimeters

OBJ. 1
5.2 (B)

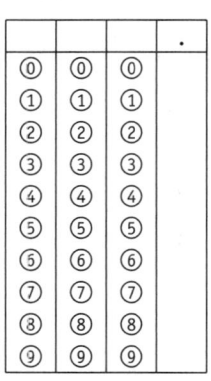

6 The graph below shows the length that a plant grew over a four-week period. According to the graph, how many additional centimeters did the plant grow from Week 2 to Week 4? Record and bubble in your answer on the bubble grid below.

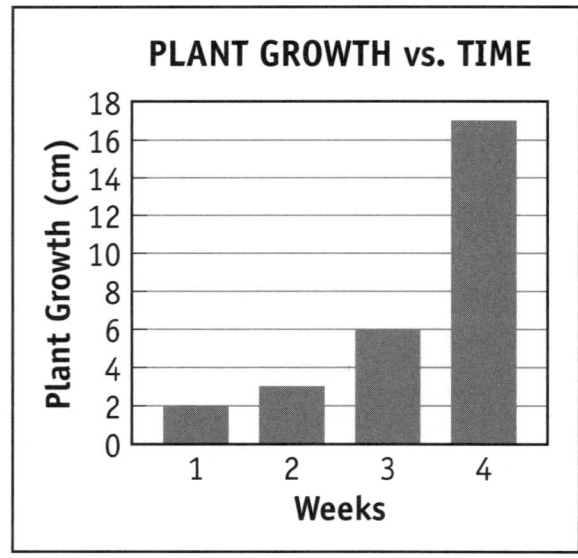

OBJ. 1
5.2 (B)

7 **A class recorded the outdoor temperature at noon on the first day of the month during their school year. When they were finished their chart appeared as follows:**

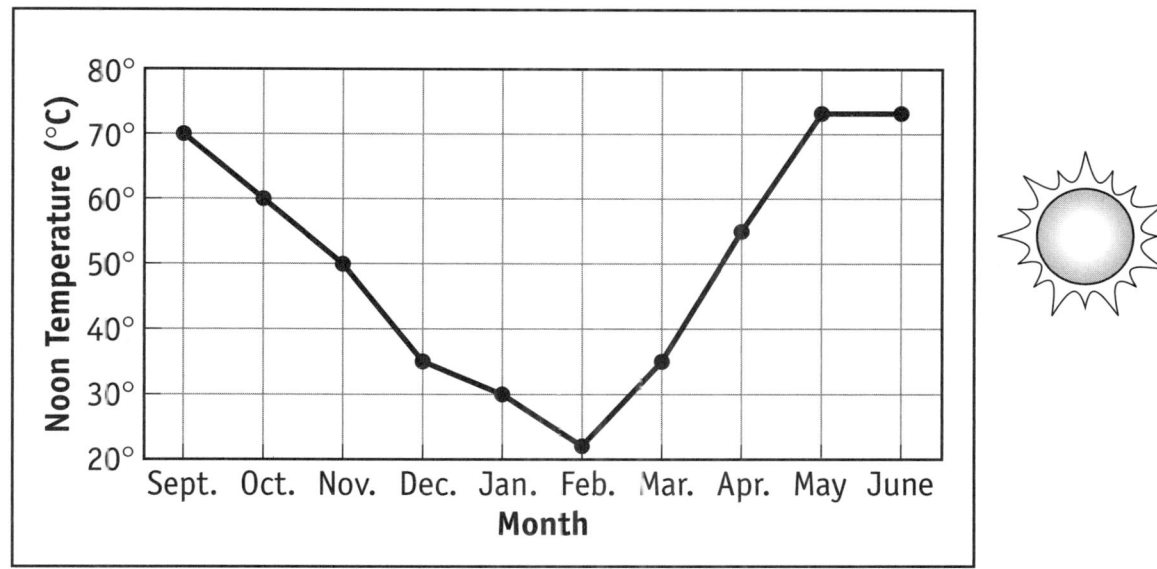

What was the temperature on the first day of March?

A 30°C

B 75°C

C 35°C

D 40°C

OBJ. 1
5.2 (E)

8 **From October to January, what happened to the temperature?**

F It decreased, then increased.

G It increased.

H It remained the same.

J It decreased.

OBJ. 1
5.2 (E)

9 **Wanda heard on the radio that a storm is expected in her area. She wants to find out how much precipitation will fall during the storm.**

What tool would be best for Wanda to use to collect her data?

A A rain gauge

B A thermometer

C A balance scale

D A microscope

OBJ. 1
5.2 (A)

10 **What will Kenesha find out by measuring the mass of her soccer ball?**

F How much space it takes up

G How much heat it has

H How much matter it contains

J How fast it is moving

◆ Examine the Question
◆ Recall What You Know
◆ Apply What You Know

OBJ. 1
5.2 (A)

11 A scientist measures the lengths of 100 tadpoles from two different ponds. The scientist then calculates the average tadpole length in each pond.

What would be the best way for the scientist to display the results?

A   Bar graph   C   Line graph
B   Circle graph   D   Map

OBJ. 1
5.2 (E)

12 The table below shows the amount of rainfall in Ft. Worth for one week.

Day	Amount of Rain
Sunday	1 cm
Monday	2 cm
Tuesday	2 cm
Wednesday	0 cm
Thursday	1 cm
Friday	0 cm
Saturday	4 cm

What was the total rainfall in Ft. Worth for that week?

F   7 cm   H   9 cm
G   8 cm   J   10 cm

OBJ. 1
5.2 (E)

A school in San Antonio had a recycling drive. They collected newspapers for five months. The results are shown in the table below:

KILOGRAMS OF NEWSPAPERS COLLECTED FOR THE RECYCLING DRIVE

Grade	October	November	December	January	February
Second Grade	42	50	42	56	53
Third Grade	27	35	47	49	59
Fourth Grade	31	40	30	40	53
Fifth Grade	35	43	40	53	47

13 Which grade showed an increase each month in the number of kilograms of newspapers it collected?

A   Second Grade
B   Third Grade
C   Fourth Grade
D   Fifth Grade

♦ Examine the Question
♦ Recall What You Know
♦ Apply What You Know

OBJ. 1
5.2 (E)

14   How long is the screw shown below? Record and bubble in your answer on the grid below.

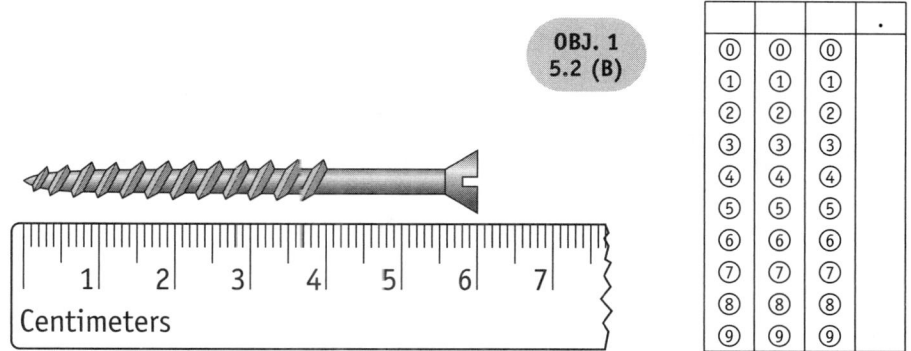

OBJ. 1
5.2 (B)

15   The table below shows the movement of a snail during a six-hour period. If the snail moves at the same pace, how far would it travel in 8 hours?

Time Traveled	Total Distance Traveled
2 hours	40 cm
4 hours	80 cm
6 hours	120 cm
8 hours	?

A   40 cm
B   80 cm

C   160 cm
D   180 cm

OBJ. 1
5.2 (E)

16   The following table shows data collected during an experiment involving the temperature recorded from Monday through Friday in El Paso, Texas.

Day	Monday	Tuesday	Wednesday	Thursday	Friday
Temperature	72° F	77° F	78° F	82° F	70° F

Which thermometer correctly shows the temperature recorded on Friday?

OBJ. 1
5.2 (C)

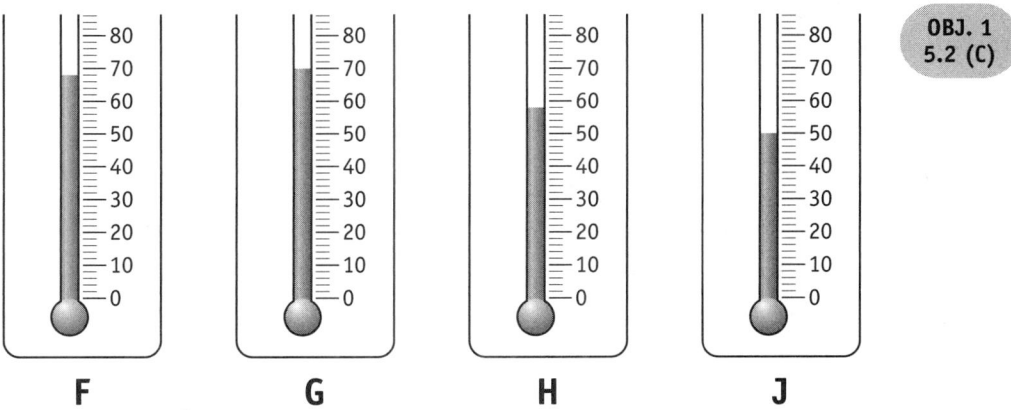

F   G   H   J

CHAPTER 3

CRITICAL THINKING IN SCIENCE

This chapter will show you how to use your critical thinking and scientific problem-solving skills to make better decisions.

— MAJOR IDEAS —

★ Scientists use **models** to represent the natural world. By examining how a model works, scientists often can develop **theories** to help explain what they have observed in nature.

★ Scientists use their observations to test and revise their theories.

★ Scientists use their critical thinking and problem-solving skills to help them draw conclusions about advertised products and services.

MODELS IN SCIENCE

Scientists use models to understand what happens in nature. They may use a model if an object is too small to be seen, like an atom, or if the object is impossible to see, like the center of Earth. They also use models to make it easier to test ideas.

A **model** is made to represent something else. It is simpler and a different size than what it represents. A model car is made of painted plastic and metal to 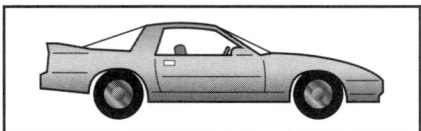 look like a real car, only it is much smaller. The model has the same shape, outside parts, and colors as the real car. Everything in the car is made to **scale** — it has the same proportions as the car it represents. On the other hand, it is not a real car in many ways. The inside of a model car is different from a real car. It has no working engine and cannot run. The purpose of a model car, however, is to show what a real car looks like from the outside. The purpose of a model often affects how it is built.

TYPES OF MODELS

Scientists often make three-dimensional, or "3-D," models. Some models use simple objects to show complex relationships. For example, a scientist might make a model of the Earth, moon, and sun by using small spheres that can be easily moved. This model can help to show the different phases of the moon, or eclipses. A model can even be just a diagram. For example, scientists may draw an atom as a nucleus surrounded by electrons.

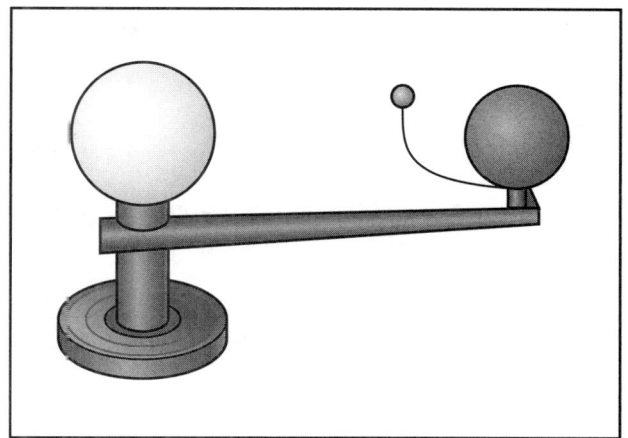

THE PURPOSE OF A MODEL

A good model helps scientists to see relationships and test ideas. More details will often improve the model. For example, scientists may change the sizes of different parts of a model or the distances between the parts. The improved model helps the scientist better understand what is happening. From the model, scientists make predictions about what will happen. Then they test their predictions to see if they come true. The more closely a model resembles what it represents, the better its predictions will usually be.

APPLYING WHAT YOU HAVE LEARNED

✦ Think of something you have learned about in science this year.

• Describe how you would make a model to show it. _____

_____

_____

_____

• Explain how your model makes this thing easier to understand. _____

_____

_____

_____

A CASE STUDY: THE THREE GORGES DAM

Scientists use models to help them predict certain processes or problems they may face before work begins on a project. For example, scientists in China were building a large dam. The Three Gorges Dam was the largest dam project in the world. They wanted to see what effects the dam would have before it was actually built. So they built a large model of the dam. The model had the same proportions and materials as the actual dam, but was much

This was the model they used on the project.

smaller. The scientists used the model to conduct several experiments before construction began. Based on their studies of this model, they were able to make improvements in the design of the dam. Below is a scale drawing used by the builders in designing the dam.

IMPROVING MODELS

Models can never be exactly the same as what they represent. They will always differ in size, materials, speed of movement or some other factors. Because of these differences, models can always be improved. The more closely a model resembles the real thing, the better it usually is.

Some questions on the **Elementary Science TAKS** may test your understanding of models. Remember, every model can be made more accurate by adding details or improving proportions. Therefore, when you examine a model, always ask yourself:

What is this model trying to show?	**How closely does this model show what it represents?**	**How might this model be improved?**

APPLYING WHAT YOU HAVE LEARNED

✦ In a model of our solar system, a tennis ball is used to represent Earth.

• What would you use to represent the sun? _____

• What would you use to represent the planet Mercury? _____

• Where would you place these objects? _____

_____

✦ What materials would you use to make a model of a volcano? _____

_____

• How would you put your model of the volcano together? _____

_____

_____

SCIENTIFIC EXPLANATIONS

You already know that scientists observe the world and ask questions. They look for patterns in the way things happen. Then they develop ideas to explain what they observe. A "big idea" in science is called a theory. A **theory** attempts to explain how and why things happen.

To come up with a theory, scientists look at data from experiments and from observing nature. Then they try to think of a logical way to explain all the data. For example, the ancient Greeks looked at the sky and saw that the moon, stars, and planets move. They thought all of these objects revolved around the Earth. However, there were some observations they could not seem to explain.

Scientists could not explain why the planets sometimes seemed to move backwards. In the 1500s, **Copernicus** studied the planets and came up with a better theory that explain all the data. He decided that the Earth moves around the sun.

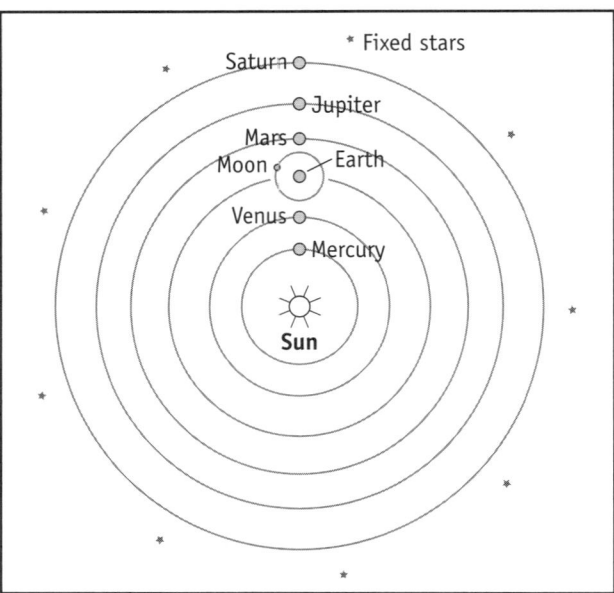

Once scientists develop a theory, they conduct experiments to test it. If the results of repeated experiments support the theory, then the theory provides a good explanation. A theory is weak if there are results that it cannot explain. Then the theory has to be changed or rejected. To decide how good a theory or hypothesis is, you must look at the data from observations and experiments. See if the data supports the theory.

LOOKING AT PRODUCTS

Manufacturers make claims for their products. Some of these claims are opinions, such as "this cereal tastes best." Other claims are factual statements that can be supported with scientific evidence. Many products make claims based on scientific research. When you see a product or service that makes a claim, use your critical thinking skills to determine if the claim is really true. For example, which information on a product's label supports its claims? Then decide if the claims are accurate.

From the information you read about a product, you should also be able to draw conclusions about when and how the product should be used. This information should also explain how to use the product safely. For example, you should not use a product that is highly flammable near a fire or a source of great heat.

Some types of products have specialized information on their labels:

FOODS

Foods are required to have nutrition labels. This information tells you how much fat, cholesterol, sodium, carbohydrates, protein and vitamins the food has.

★ **Serving Size.** Serving sizes are given in common household as well as metric measurements. For example, the serving size of this chicken noodle soup is $\frac{1}{2}$ a cup.

★ **Servings Per Container.** This tells you how many servings you can expect to get. In this example, there are about two and a half servings in the package.

★ **Amount Per Serving** comes next. There are 60 calories in one serving of chicken noodle soup.

★ **Nutrients** tell you how much of each nutrient there is in a serving. It's hard to know if that amount is a lot or a little. To make your job easier, the Nutrition Facts label includes **% Daily Value**. For example, if a food contains 8 grams

Chicken Noodle Soup	
Nutrition Facts	
Serving Size 1/2 cup (120 ml) condensed soup	
Servings Per Container about 2.5	
Amount Per Serving	
Calories 60	Calories from Fat 15
	% Daily Value*
Total Fat 1.5g	2%
Saturated Fat 0.5g	3%
Trans Fat 0g	
Cholesterol 15mg	
Sodium 890mg	37%
Total Carbohydrate 8g	3%
Dietary Fiber 1g	4%
Sugars 1g	
Protein 3g	
Vitamin A 4% • Vitamin C 0%	
Calcium 0% • Iron 2%	
*Percent Daily Values are based on a 2,000 calorie diet. Your daily values may be higher or lower depending on your calorie needs.	

of carbohydrates, the % Daily Value is the clue. It tells you that one serving has 3% of the carbohydrates you need that day. Daily Values are based on a daily diet of 2,000 calories.

★ **Ingredients** are required on labels of all foods with more than one ingredient. Ingredients are listed in order by weight, from most to least. If you have food allergies, the ingredients list can help you identify foods that might be a problem for you.

SKIN-CARE PRODUCTS

Many skin care products also provide protection from the sun's harmful ultraviolet rays, which may cause skin cancer. "**Sun protection factor**" or **SPF** is a measure of a sunscreen's ability to protect the skin and to prevent sunburn. The SPF scale for sunscreen ranges from 2 to 52. The number states how long you can be in the sun before your skin burns.

FSC
Don't burn, get FSC today!

Fun Sun Cream

• Great for all outdoor activities
• Contains the finest ingredients
• No mess! No waste!
• Handy container goes anywhere   *Money back guarantee*
• Sun protection factor (SPF) 30

OVER-THE-COUNTER MEDICINES

Medicines that you buy in the store or pharmacy will give directions on how they should be used. They will also list possible side-effects to watch out for. Medicines list both their active and inactive ingredients.

APPLYING WHAT YOU HAVE LEARNED

✦ Mary Ann went to the store to do her weekly shopping. She picked up half a gallon of ice cream with the information on its nutrition label on the right. Which information should Mary Ann read carefully to decide if this ice cream is a healthy food for her family to eat?

_____

_____

_____

✦ Why is the "SPF number" important when making a decision about buying a skincare product? _____

_____

_____

Nutrition Facts

Serving Size 1 order (211g)

Amount Per Serving

Calories 500   Calories from Fat 260

	% Daily Value*
Total Fat 29g	45%
Saturated Fat 10g	50%
Trans Fat	
Cholesterol 325mg	108%
Sodium 1600mg	67%
Total Carbohydrate 33g	11%
Dietary Fiber 1g	4%
Sugars 2g	
Protein 25g	

Vitamin A	10%	• Vitamin C	0%
Calcium	30%	• Iron	20%

*Percent Daily Values are based on a 2,000 calorie diet. Your daily values may be higher or lower depending on your calorie needs.

WHAT YOU SHOULD KNOW

A. You should understand that a scientific model represents processes or objects and that any model might be improved.

B. You should be able to determine the strengths and weaknesses of a scientific explanation based on how well it is supported by scientific data and observations.

C. You should be able to draw conclusions from information about products and services.

CHAPTER STUDY CARDS

Models

★ Scientists use a model to represent something else, like a process or object.

★ Models have different purposes. They help scientists to test their ideas and make predictions about what will happen.

★ Models are usually made to scale.

★ To improve a model you can often:
  • change the size and location of its parts.
  • make it more detailed, so that it more closely resembles what it represents.

Evaluating Product Information

★ To evaluate a product, read the information on its label or advertising carefully.

★ Decide which information on the label or advertisement supports a product's claims.

★ Carefully follow all information about how to use the product.

★ Each food product has a **nutrition label** that provides important information about its ingredients.

★ Skin care products have a sun protection factor (SPF) which tells how much protection it gives skin from the sun's rays

CHECKING YOUR UNDERSTANDING

1 **Cereals are made of different ingredients. The diagram to the right shows some of the ingredients in a box of Sam's Breakfast Cereal.**

A person should not eat this cereal who is allergic to —

A nuts

B skimmed milk

C wheat

D broccoli

OBJ. 1
5.3 (B)

Nutrition Facts

Ingredients:
Whole wheat, wheat bran, oats, raisins, rice, sugar, dates, dried fruit.

Sam's
Breakfast
Cereal

This question examines your ability to draw conclusions from information about a product. The **Nutrition Facts** label gives the ingredients about a food product so a person allergic to one of its ingredients can avoid that food. Since this cereal has wheat as one of its ingredients, a person allergic to wheat should avoid eating it. Therefore, the best answer is **Choice C**.

Now try answering some additional questions on your own

2 Tyra has made a model of the solar system in her garage. A lamp without a shade is placed in the center of her garage to represent the sun. A basketball next to the garage wall is used to represent the Earth. What is one way she could make her model more accurate?

  F Use a tiny pin head to represent the Earth

  G Make the size of the sun smaller

  H Use a light bulb for the Earth

  J Move the basketball to the center of the garage

OBJ. 1
5.3 (C)

♦ **Examine the Question**
♦ **Recall What You Know**
♦ **Apply What You Know**

3 Which conclusion could be reasonably made about Texas Cola?

  A It is made up of mostly sugar.

  B The largest ingredient is water.

  C It has no calories.

  D The can provides 2 servings.

OBJ. 1
5.3 (B)

4 Carlos is allergic to certain grains. Which of the following would be the fastest way for Carlos to learn if he is allergic to a candy bar?

  F Ask the supermarket clerk

  G Check its "Nutrition Facts"

  H Speak to a nutritionist

  J Write to the candy bar company

OBJ. 1
5.3 (B)

5 Lakeisha likes to eat a snack that claims it is "low in fat." How can she check if the claim is accurate?

  A Check the bar code on the box

  B Look at the "Nutrition Facts" on the box

  C Examine the weight of the box

  D Observe the number of servings in the box

OBJ. 1
5.3 (B)

6 The illustration on the right shows the "Nutrition Facts" for a cheeseburger at a popular fast food restaurant. How many calories are there in one serving of this cheeseburger?

F 29
G 211
H 350
J 500

OBJ. 1
5.3 (B)

Nutrition Facts
Serving Size 1 order (133g)

Amount Per Serving		
Calories 360	Calories from Fat 160	
		% Daily Value*
Total Fat 17g		26%
Saturated Fat 0g		40%
Trans Fat 1g		
Cholesterol 50mg		17%
Sodium 790mg		33%
Total Carbohydrate 31g		10%
Dietary Fiber 2g		8%
Sugars 6g		
Protein 19g		
Vitamin A	6% • Vitamin C	2%
Calcium	15% • Iron	20%

*Percent Daily Values are based on a 2,000 calorie diet. Your daily values may be higher or lower depending on your calorie needs.

7 How much sodium is there in one serving of this cheeseburger?

A 67 mg
B 160 mg
C 500 mg
D 790 mg

OBJ. 1
5.3 (B)

8 Which model best shows the effects of wind erosion on the Earth's surface?

F

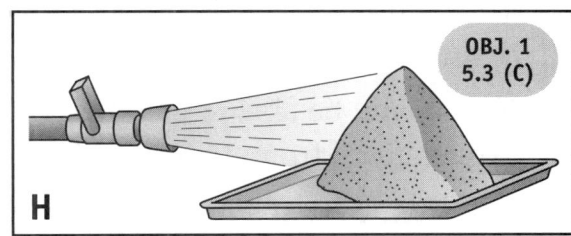

H

OBJ. 1
5.3 (C)

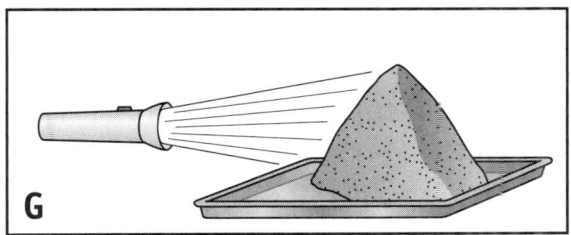

G

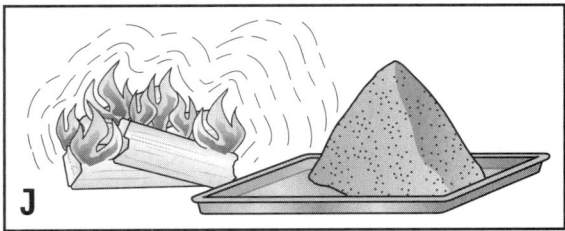

J

9 Which of these statements tells how well this jar of skin moisturizer helps to prevent sun damage?

A Keeps skin young-looking
B Contains 3 moisturizers
C All natural ingredients
D Has an SPF of 20

OBJ. 1
5.3 (B)

Joan's
Skin Moisturizer
• Keeps skin young looking
• Contains 3 moisturizers
• Hypoallergenic
• All natural ingredients
• Sun protection factor (SPF) 20

CHECKLIST OF OBJECTIVES IN THIS UNIT

*At the end of each content unit you will find a **Checklist of Objectives** like the one below. The purpose of these checklists is to help you mentally review the major objectives examined in the unit before moving on to the next unit.*

Directions. Now that you have completed this unit, place a check (✔) next to those objectives you understand. If you are having trouble recalling information about a particular objective, review the chapter listed in the accompanying brackets.

☐ You should be able to demonstrate safe practices during field and laboratory investigations. [**Chapter 1**]

☐ You should be able to plan and implement descriptive and simple experimental investigations including asking well-defined questions, formulating testable hypotheses, and selecting and using equipment and technology. [**Chapter 1**]

☐ You should be able to collect information by observing and measuring. [**Chapter 1**]

☐ You should be able to analyze and interpret information to construct reasonable explanations from direct and indirect evidence. [**Chapter 1**]

☐ You should be able to communicate valid conclusions. [**Chapter 1**]

☐ You should be able to construct simple graphs, tables, maps, and charts using tools to organize, examine, and evaluate information. [**Chapter 2**]

☐ You should be able to analyze and review scientific explanations, including hypotheses and theories, as to their strengths and weaknesses using scientific evidence and information. [**Chapter 2**]

☐ You should be able to draw inferences based on information for products and services. [**Chapter 3**]

☐ You should be able to represent the natural world using models and be able to identify their limitations. [**Chapter 3**]

☐ You should be able to collect and analyze information using tools, including calculators, microscopes, hand lenses, rulers, thermometers, compasses, balances, meter sticks, timing devices, magnets, collecting nets, and safety goggles. [**Chapters 1 and 2**]

UNIT 2

LIFE SCIENCES

In this unit, you will review what you need to know for the **Elementary Science TAKS** about the **life sciences** — the study of living things.

You will learn how living things meet their needs, how plants and animals differ, and how living things interact with their environment in ecosystems.

Flamingos interacting with their environment

★ Chapter 4: Plants and Animals

In this chapter, you will learn the four basic needs of all living things. You will also learn how plants are different from animals, and how both plants and animals meet their needs.

★ Chapter 5: Ecosystems

This chapter looks at how different types of plants and animals live together and interact with their environment in ecosystems. You will also learn how environmental changes affect living things, and how living things affect their environment.

★ Chapter 6: Inherited Traits and Learned Behavior

In this chapter, you will learn the difference between learned behavior and inherited traits.

CHAPTER 4

PLANTS AND ANIMALS

In this chapter, you will learn the main characteristics of plants and animals.

— MAJOR IDEAS —

★ All living things need air (dissolved gases), water, food and space. All living things depend on each other and on the environment.

★ Plants are able to make their own food using the energy of sunlight, while animals are unable to make their own food.

★ Plants are unable to move from one place to another.

★ Animals can move from one place to another. They use their senses to guide their movement.

★ Plants and animals each go through their own unique life cycles.

THE NEEDS OF LIVING THINGS

All living things have certain basic needs: air (or gases), water, food and space.

APPLYING WHAT YOU HAVE LEARNED

Describe how you go about meeting your needs for:

◆ Air: _____

◆ Water: _____

◆ Food: _____

◆ Space / Shelter: _____

PLANTS AND ANIMALS

Living things, or **organisms**, have different characteristics to help them meet their needs. Two types of living things with different characteristics are plants and animals.

APPLYING WHAT YOU HAVE LEARNED

Plants and animals are all around you. List some of the characteristics you know for plants and animals:

PLANTS	ANIMALS
• _____	• _____
• _____	• _____
• _____	• _____
• _____	• _____

Scientists have investigated living things to see what the differences between plants and animals are. Here are some of the things they have found out:

PLANTS

Plants are green. Their cells contain a chemical called chlorophyll. **Chlorophyll** helps plants turn energy from the sun into food. Because plants can make their own food, they are called **producers**. When scientists look at plants under a microscope, they see that plant cells are surrounded by stiff cell walls. These stiff cell walls help plants stand. However, because of their cell walls, plants cannot move about from place to place.

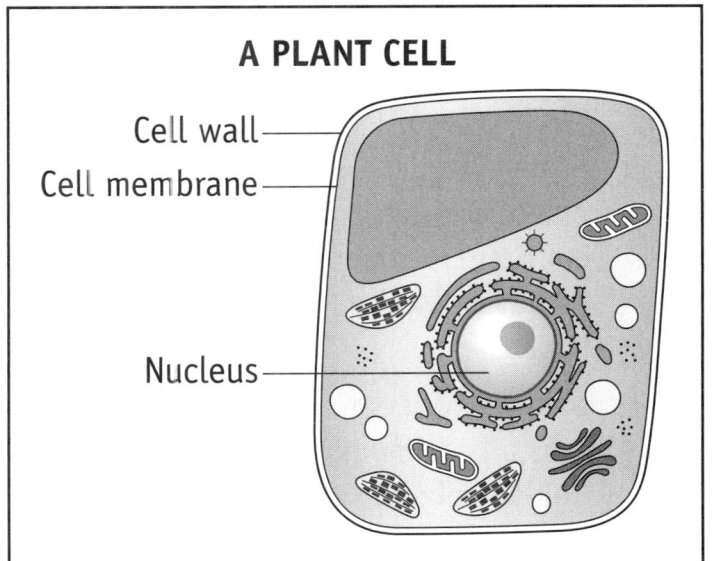

A PLANT CELL

Cell wall

Cell membrane

Nucleus

Scientists also look at animal cells under a microscope. They have found that animal cells do not have chlorophyll. They are unable to make their own food. To get food, animals must eat plants or other animals. Animal cells also do not have cell walls, so animals can move around.

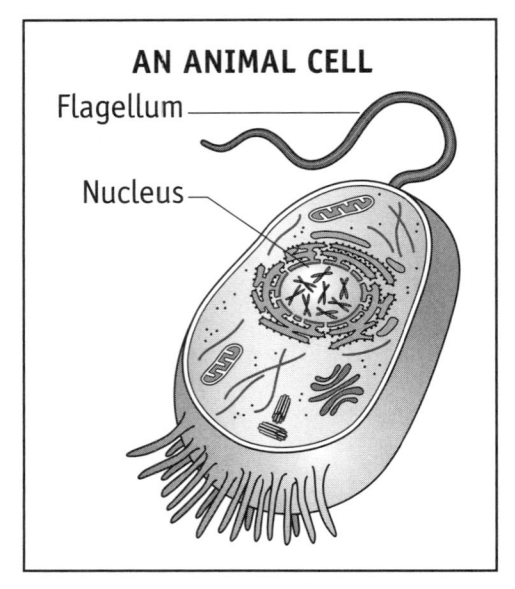

AN ANIMAL CELL

Flagellum

Nucleus

HOW PLANTS
MEET THEIR NEEDS

There are many kinds of plants. Each has its own special characteristics to help it meet its basic needs. Plants have three parts: a *root*, *stem*, and *leaves*. Each part of the plant helps it to meet its basic needs.

★ **Leaves.** Leaves make food from the sun's energy. This process is called **photosynthesis**. Tiny pores in the bottom of the leaves absorb carbon dioxide from the air and give out oxygen. In the leaves, sunlight is mixed with carbon dioxide to produce a type of sugar. The plant uses this sugar as food. Photosynthesis also gives off water and oxygen.

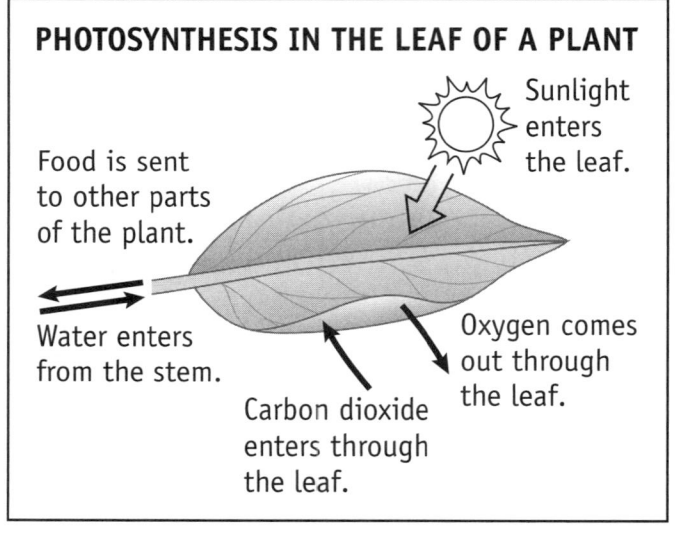

PHOTOSYNTHESIS IN THE LEAF OF A PLANT

Sunlight enters the leaf.

Food is sent to other parts of the plant.

Water enters from the stem.

Carbon dioxide enters through the leaf.

Oxygen comes out through the leaf.

★ **Stems.** Stems are the main body of the plant. They support its leaves and flowers. Stems also move water and minerals to the leaves, and food from the leaves to the roots.

★ **Roots.** Roots are the part of the plant typically found below the surface. They hold the plant in the ground, and absorb water and minerals from the soil.

APPLYING WHAT YOU HAVE LEARNED

Answer the following questions based on the diagram on page 54:

◆   How do plants get food? _____
_____

◆   How do plants stay in the same place? _____

◆   How do plants get water? _____

THE LIFE CYCLE OF PLANTS

Do you look the same as you did five years ago? Of course not! As you live, you change. All living things go through steps known as **life cycles**. All living things begin life, grow, age and eventually die. As living things age, they often go through changes. Here is the life cycle of a typical flowering plant:

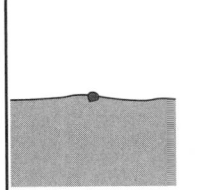 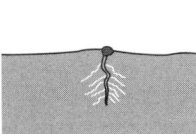  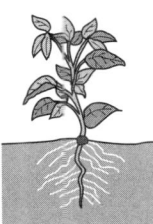  

A seed lands on the ground.	In good conditions, the seed germinates. Then it grows roots that push into the soil.	The seed sprouts and becomes a seedling. The seedling uses the food in the seed.	The seedling begins to grow. Soon it grows into a plant.	The plant develops flowers. Pollen from another flower lands on the flower's stigma.	A new seed now develops in the plant.

Spreading Seeds. Plants cannot move around. Seeds need to be spread away from their parent plant if they are to survive. Seeds that grow too close to a parent plant will compete with the parent plant for food. Instead, many seeds are blown by the wind. Plants often have a fruit which is eaten by animals. When the seed is passed out or discarded by the animal, it grows in another place.

APPLYING WHAT YOU HAVE LEARNED

Carefully examine the illustration to the right. Reorder the different plants in the order of their development in the plant life cycle.

◆ Step 1: _____

◆ Step 2: _____

◆ Step 3: _____

◆ Step 4: _____

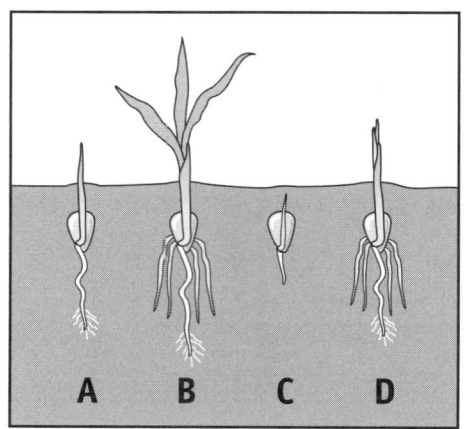

HOW ANIMALS MEET THEIR NEEDS

All animals have something in common. They cannot make their own food. In order to obtain energy, they must eat plants or other animals. Because they depend on eating other living things to survive, animals are called **consumers**.

Different kinds of animals have special characteristics that help them meet their needs in the environment.

SALTWATER ENVIRONMENT

Fish, dolphins, jellyfish, octopuses and many other animals live in a saltwater environment.

Fish, for example, meet their need for oxygen by swimming through the water. Their **gills** filter the water and absorb oxygen from the water. Some fish eat smaller fish for food. They have sharp teeth to help them tear apart the food they eat. Many fish have flat bodies that allow them to rest on the seafloor. Other saltwater animals scrape the seafloor for something to eat.

FRESHWATER ENVIRONMENT

Oceans make up 97% of the Earth's water. The remaining 3% is freshwater. Most of the Earth's fresh water is stored as ice or groundwater. The remaining fresh water, making up less than 1% of the total, is found in lakes, rivers and wetlands. Fish, tadpoles and other freshwater animals live in these lakes, rivers and streams. They eat other animals and plants. Freshwater fish absorb oxygen through their gills just like saltwater fish. However, they cannot live in salt water.

LAND ANIMALS

There are many types of land environments — deserts, tropical rain forests, grasslands, and very cold places. Different types of animals live in each of these environments. They have special characterisitics to help them meet their needs. All land animals, however, breathe **oxygen** from the air around them. They drink water, and move from place to place to find food. Usually they have limbs to help them move. This helps them to meet their basic needs.

APPLYING WHAT YOU HAVE LEARNED

Choose an animal you know. Then explain how that animal meets its basic needs:

For Space?	
For Oxygen?	
For Food?	
For Water?	

Unlike plants, animals can move about to find food. They also use their different senses to help them move and find food, air, water and shelter. Some simple animals may have only a few senses, but most animals have five senses just like we do.

APPLYING WHAT YOU HAVE LEARNED

Do you know your five senses? How do you use each of them to survive?

Sight:	
Smell:	
Hearing:	
Touch:	
Taste:	

THE LIFE CYCLE OF ANIMALS

Like plants, animals also go through life cycles. Reptiles and birds are born from eggs. Mammals have live births. Some animals go through special stages and actually change from one form to a completely different form. For example, a frog begins as an egg. Eventually, the egg will hatch and form a tadpole. After several weeks, the tadpole will develop tiny legs and arms. Then the tadpole grows into a young frog.

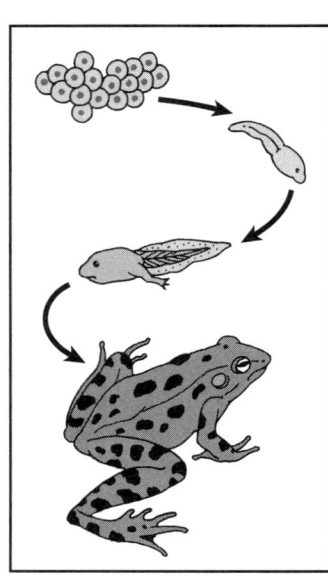

Some insects go through a special process, known as **metamorphosis**. They change their physical form after birth.

For example, a butterfly lays an egg. Out of the egg comes a **larva** (*or caterpillar*). The larva wraps itself up in a **pupa** (*or cocoon*). After a period of time, an adult butterfly emerges from the pupa.

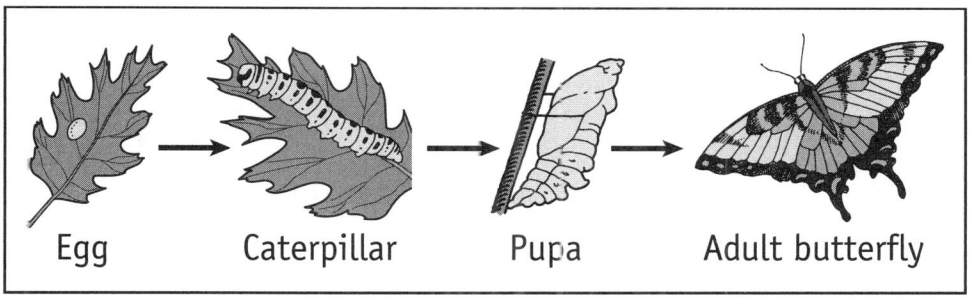

Egg   Caterpillar   Pupa   Adult butterfly

APPLYING WHAT YOU HAVE LEARNED

How do the characteristics of plants and animals differ?

Plants	Animals

WHAT YOU SHOULD KNOW

+ You should know that all living things need certain items to survive: air (or dissolved gases), water, food, and space.

+ You should know that plants cannot move from place to place. Plants are able to make their own food using the energy of sunlight in a process known as **photosynthesis**. Animals cannot make their own food.

+ You should know that animals can move from one place to another. They use their different senses to guide their movements.

+ You should know that different types of plants and animals each go through their own unique life cycles.

CHAPTER STUDY CARDS

Plants

★ **Photosynthesis**. Plants make their own food out of sunlight through a process called photosynthesis.

★ Plants have three main parts:

  • **Roots:** Anchor the plant in the ground.
  • **Leaves:** Make food from the sun's energy
  • **Stems:** Support its leaves and flowers.

★ Plants go through life cycles: a seed, the seed germinates, a seedling, then a mature plant.

★ Plants use carbon dioxide and produce oxygen.

Animals

★ Animals cannot make their own food. They must eat plants and/or other animals.

★ Unlike plants, animals move from place to place to meet their needs.

★ Animals use their senses — sight, hearing, smell, touch, and taste — to help them meet their needs of survival.

★ **Life Cycles**. Animals go through life cycles. In **metamorphosis**, an animal changes its form.

★ Animals breathe in oxygen and exhale carbon dioxide.

CHECKING YOUR UNDERSTANDING

Examine the following illustrations:

  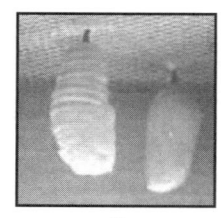 

     1   2   3   4

1   Which is the correct order of development for this butterfly?

A $4 \rightarrow 2 \rightarrow 1 \rightarrow 3$

B $1 \rightarrow 4 \rightarrow 2 \rightarrow 3$

C $3 \rightarrow 2 \rightarrow 4 \rightarrow 1$

D $1 \rightarrow 3 \rightarrow 4 \rightarrow 2$

OBJ. 2
5.6 (C)

HINT

To answer this question correctly, you must know about the metamorphysis of a butterfly in its lifetime. When its egg hatches, a larva comes out. The larva later becomes a pupa. Out of the pupa comes a beautiful butterfly. If the pictures were arranged in their order of development, the order would be 3 – 2 – 4 – 1. Thus, **Choice C** is correct.

Now try answering some additional questions on your own:

CHARACTERISTICS

Producers	Organisms (*living things*) that produce their own energy through photosynthesis
Consumers	Organisms that must consume other organisms for energy

2 **Which group of organisms contains only consumers?**

  F Wheat, snakes, mice
  G Acorns, squirrels, owls
  H Grass, rabbits, foxes
  J Rats, chipmunks, owls

OBJ. 2
2.9 (A)

3 **Owls eat mice and other small animals to survive. Owls have large eyes to help them hunt for their next meal in the dim light of night. Which of the owl's senses is most useful as it hunts?**

  A Taste
  B Touch
  C Smell
  D Sight

◆ **Examine the Question**
◆ **Recall What You Know**
◆ **Apply What You Know**

OBJ. 2
2.9 (A)

4 **An anteater survives mainly by finding and eating ants. It hunts by sticking its long snout (nose) in the ground to find ants. Which of the anteater's senses is most useful in seeking out food?**

  F Its ability to see
  G Its sharp hearing
  H Its sense of smell
  J Its sense of touch

OBJ. 2
2.9 (A)

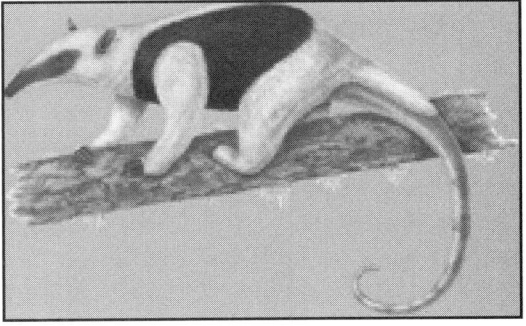

Anteater

5 **When cows graze on a field they can often be seen eating grass. What basic need do the cows meet by eating the grass?**

  A Water
  B Space
  C Food
  D Oxygen

OBJ. 2
2.9 (A)

Cows grazing in a pasture

6 **In order for an oak tree to make its own food, it leaves must absorb —**

F water
G sunlight
H minerals
J oxygen

◆ Examine the Question
◆ Recall What You Know
◆ Apply What You Know

OBJ. 2
2.9 (B)

Examine the following diagram

ANIMALS
- consume food
- produce carbon dioxide

BOTH
- _____

PLANTS
- produce food
- produce oxygen

7 **What best completes the blank line?**

A Stationary
B Moves about freely

C Needs water
D Has stiff cell walls

OBJ. 2
2.9 (A)

8 **All of the following are what a fish needs to live in saltwater or fresh water except —**

F food
G sand

H water
J oxygen

OBJ. 2
2.9 (B)

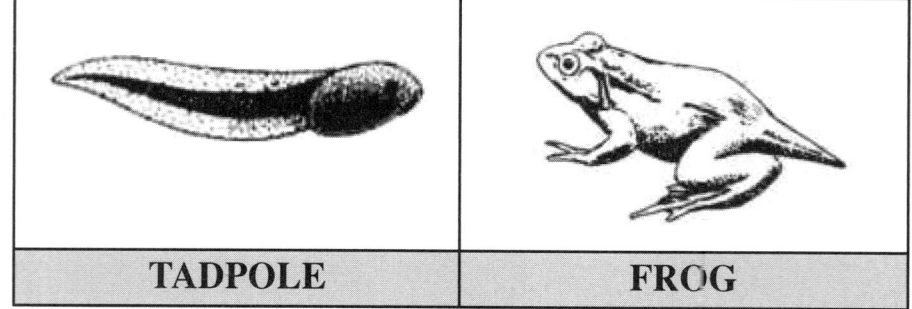

TADPOLE   FROG

9 **Which of the following pairs is most like the pair above?**

A Pupa – butterfly
B Worm — snake

C Branch — tree
D Egg — seed

OBJ. 2
5.6 (C)

10 **Fish that swim in the ocean, a pet dog, and an owl are alike in important ways. One of the ways this group is alike is that all three have —**

F legs
G hair

H gills
J eyes

OBJ. 2
2.9 (A)

11 **Green plants are important to animals because plants —**

    A consume food and give off oxygen

    B consume food and give off carbon dioxide

    C produce food and give off oxygen

    D produce food and give off carbon dioxide

OBJ. 2
2.9 (B)

12 **The diagram shows a bean seed in a container of soil. The seed is starting to germinate. The first part to come out is labeled X. What is the function of X?**

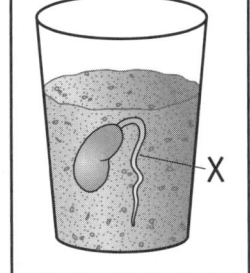

    F to make food

    G to absorb water

    H to produce other seeds

    J to release oxygen

OBJ. 2
5.6 (C)

13 **Megan and Maria want to grow carrots in their garden. What do they need to bury in the soil?**

    A A carrot seed

    B The stem of a carrot

    C A carrot flower

    D A carrot leaf

OBJ. 2
2.9 (A)

14 **Which of the following do plants use to make food?**

    F Water

    G Oxygen

    H Sugar

    J Sunlight

OBJ. 2
2.9 (A)

15 **In order to survive, every animal must have —**

    A roots, leaves, and stems   **C** eyes, nose, and ears

    B food, water, and oxygen   **D** light, soil, and nutrients

OBJ. 2
2.9 (B)

16 **A student decides to conduct an experiment about seed germination. He plants several seeds. Which part of the seed will grow first if it germinates?**

    F Leaf   **H** Root

    G Stem   **J** Flower

OBJ. 2
5.6 (C)

17 The pictures below show the main stages in the life cycle of a fly. Which group shows the life cycle of the fly in the right order?

OBJ. 2
5.6 (C)

A

B

C

D

18 Which of the items listed below is a main function of this plant's roots?

OBJ. 2
2.9 (A)

F Making seeds
G Producing pollen
H Absorbing nutrients
J Storing oxygen

19 Plant life produces seeds. Often these seeds are spread by the wind. Which of the seed types below is best suited to be spread by the action of the wind?

 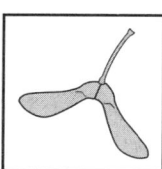 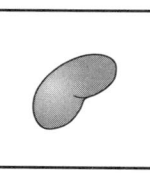 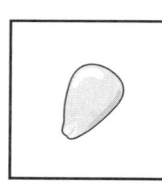

OBJ. 2
2.9 (A)

A   B   C   D

20 Which basic need do birds meet by building nests in trees?

F Water
G Food
H Shelter
J Air

OBJ. 2
3.8 (D)

◆ Examine the Question
◆ Recall What You Know
◆ Apply What You Know

CHAPTER 5

ECOSYSTEMS

In this chapter, you will learn how different types of living things survive together and interact in ecosystems.

— MAJOR IDEAS —

★ An **ecosystem** is made up of all the living and nonliving things in a particular area. The living things in an ecosystem depend on both their physical environment and one another in order to survive.

★ Similar living things in an ecosystem compete with each other for resources, such as oxygen, water, food or space.

★ Living things in an ecosystem have **adaptive characteristics**. These characteristics help them to survive and reproduce.

★ Living things in an ecosystem often modify their physical environment.

★ Changes in an environment can cause some types of living things to die out.

WHAT IS AN ECOSYSTEM?

A **system** is a group of things or parts that act together. An **ecosystem** is a system made up of all the living and nonliving things in a particular area. Every person, animal, plant, stream and area of land or water belongs to one or more ecosystems. Because they are in the same area, they affect each other in different ways.

A pond's animals and plants form an ecosystem

A small pond provides a good example of an **ecosystem**. The water, air, sunlight, and the mud at the bottom of the pond form the **physical environment** of the pond. The moss, pond grass, and small green algae in or around the pond are forms of plant life in this ecosystem. Insects, snails and other animals living in the pond eat some of these plants to survive. Fish eat snails, insects or smaller fish in the pond, while frogs in the pond eat insects found there.

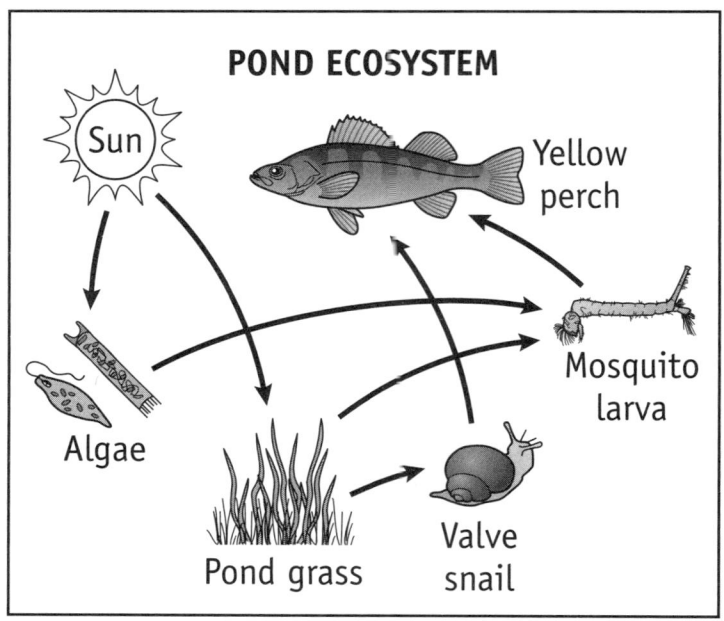

Each of the living things in the pond has its own **habitat**, or special place. Moss grows on the rocks next to the pond. Fish swim in the pond. Frogs jump in and out of the pond. Each of these are habitats in the pond's ecosystem. The fish, frogs, and insects all leave behind wastes. Snails, insects, fungi, and bacteria in and around the pond break down these wastes. Left undisturbed, a pond can continue like this for hundreds or even thousands of years. The different living things in this pond **ecosystem** live in a balanced relationship.

APPLYING WHAT YOU HAVE LEARNED

Answer the following questions about an ecosystem:

◆ How do the plants in a pond's ecosystem depend on their physical environment to meet their basic needs? _____

_____

◆ How do the animals in this ecosystem depend on other living things to meet their basic needs? _____

_____

◆ What are some of the habitats of the animals in this pond ecosystem? ___

_____

_____

THE PHYSICAL ENVIRONMENT

Living things in an ecosystem must be able to survive and reproduce in that ecosystem's nonliving, or **physical environment**. Each type of plant or animal in an ecosystem has **adaptive characteristics**. These characteristics help it to survive and reproduce in that physical environment.

ALLIGATORS

Alligators are reptiles. They have long tails which help them to swim. They store fat that can be used in the winter. Alligators have long snouts and smooth skin. These characteristics aid the alligator in swimming. Alligators have small legs so they can walk on land. They are equally at home on land or in the water, which is perfect for the swamplands they inhabit.

How does an alligator's long tail help it to survive?

GIRAFFES

Giraffes live in the grassy plains of Africa. They have excellent eyesight and can see for many miles. Giraffes eat up to 75 pounds of food each day. Their long necks allow them to meet their food needs by eating foods that are too high for most other animals. They can eat the leaves of trees with their long necks. Scientists believe a giraffe's skin pattern makes them look like tall trees. This helps them hide from other animals.

Why is a giraffe's long neck important to its survival?

CACTI

A cactus is a plant that can live in desert conditions. To survive in dry conditions, cacti have special ways of storing water. The roots of a cactus go deep under the ground. Its net of side roots collect every drop of water falling on the ground. The water is stored in the fleshy stem of the cactus. The cactus swells up in size after a rainfall. It then gets thinner as it uses up its water supply. To protect itself from animals, it has sharp needles on its outer skin. These needles or spines also reduce the amount of water that evaporates under the hot desert sun.

BEARS

There are many types of bears. They are found in many different kinds of ecosystems. Each type of bear has special adaptive characteristics that help it to survive in that kind of physical environment. For example, polar bears have black skin and white fur. Their skin can trap 90% of the energy from sunlight. This helps polar bears survive in cold climates.

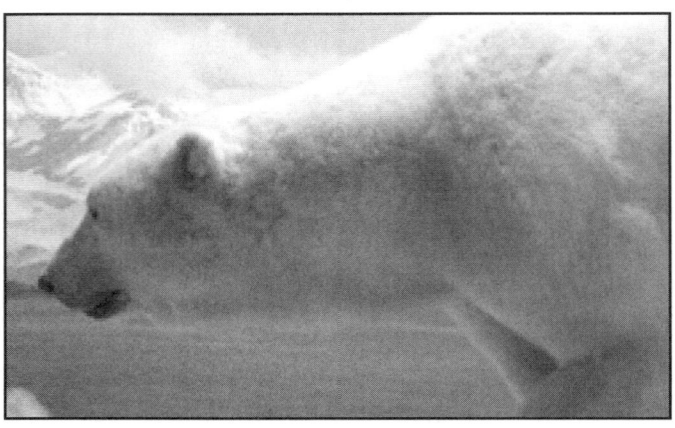

Why do bears have thick fur?

APPLYING WHAT YOU HAVE LEARNED

◆ What adaptive characteristics help cacti to survive in dry climates? _____

_____

_____

◆ A **camel** is a large mammal with one or two humps of body fat. They can store water in their blood and can live without drinking for two weeks. They can exist without food for a month. Camels are able to withstand changes in body temperature. Their thick coat of camel hair is able to reflect sunlight.

How do these physical characteristics help a camel live in a desert climate?

_____

_____

_____

LIVING THINGS CAN MODIFY THEIR ENVIRONMENT

Just as plants and animals are affected by their environment, they also affect their environment. Many animals, for example, change their physical environment to meet their needs.

MOLES

These small mammals dig tunnels underground to make their homes. Their bodies are covered in fine, dark fur. Their eyes and ears are extremely small so that they are not filled with sand while digging. Mole tunnels provide shelter and safety from larger animals above ground. Moles eat worms and insects they find as they dig their tunnels.

A mole digs a tunnel.

BEAVERS

Beavers are small mammals that live in rivers and streams. Beavers are great architects that compare to humans in their ability to change the area they live in. They build dams with tree trunks and sticks that they cut with their sharp teeth. The dam created by a beaver often raises the water level. This creates an artifical pond surrounded by fast-flowing water. Beavers make their home in this pond, where they are protected from other animals.

A beaver using its sharp teeth.

BIRDS

Birds create nests to hold their eggs. Usually a bird nest is made of twigs, grass, and leaves. Most birds build their nests in trees, but others, like eagles, build their nests on rocky cliffs. Each type of bird has its own kind of nest with its own special shape and materials.

APPLYING WHAT YOU HAVE LEARNED

◆ What basic needs do moles meet by digging tunnels in the soil? _____

_____

_____

◆ Why do beavers build dams to change their environment? _____

_____

_____

APPLYING WHAT YOU HAVE LEARNED

✦ Ants live in large colonies under-ground. These colonies can have from a few dozen workers to many thousands. Worker ants spend the first few days of their lives caring for the queen ant. After that they gradu-ate to digging tunnels and building large underground rooms to store their eggs and food. Ants often carry

Worker ants carry leaves.

dirt outside the tunnel and leave it at the entrance, forming an ant hill.

What basic needs do these underground rooms serve? _____

_____

_____

THE LIVING ENVIRONMENT

The plants and animals in an ecosystem also affect each other. Each type of animal or plant in the ecosystem has adaptive characteristics to help it live with the other living things in that ecosystem.

COMPETITION

Competition is the struggle for survival. Similar types of living things in the same ecosystem will compete for food, water and space. For example, both antelopes and zebras eat grass on the African plains. If there are more zebras and they eat more grass, there will be less grass for the antelopes. If there is less grass for the antelopes to eat, their numbers will decrease.

Antelopes and zebras compete for grass.

PREDATORS AND PREY

Many types of animals in an ecosystem live by eating plants. However, other animals survive by eating animals in the ecosystem. For example, lions on the African plains cannot live by eating grass. They can only survive by eating other animals. An animal that lives by hunting and eating other animals is called a **predator**. The animal that is hunted is known as its **prey**.

A lynx (predator) chases a rabbit (prey)

PREDATORS

Predators often have special **adaptive characteristics** to help them hunt better. These special characteristics include speed, the ability to sneak up while approaching prey, and strong senses of smell, sight and hearing.

★ **Lions** are large members of the cat family. They often weigh between 150 and 200 kg (330-600 pounds). Their large size, powerful claws, and sharp teeth help lions hunt for antelopes, zebras, and other large animals.

★ **Frogs** have long, sticky tongues. They can also jump far and fast. This helps them to catch flying insects in their mouths.

★ **Sharks** have sleek bodies so they can swim fast to catch other fish. Their keen sense of smell helps them to find their prey. Sharks have many small, sharp teeth in rows. Some sharks have thousands of teeth. These teeth face inward to make it difficult for prey, once caught, to break away.

★ **Owls** have unusually large eyes to help them see their prey at night. They have excellent hearing, while their powerful claws and sharp beak allow them to attack small mammals and birds.

PREY

Prey also have **special characteristics** to help them survive. For example, many types of prey have eyes on the sides of their head. This helps them to see if predators are coming from any direction.

For example, giraffes can see preda-tors coming from a great distance because of their long necks. Horses, antelopes and zebras can run very fast. This helps them to escape when they are in danger of being attacked. Other important **adaptive characteristics** that help prey include the ability to hide from a predator, and a good sense of smell to warn them when a predator approaches. Some prey can shoot pointy spikes or spray poison when they are bitten.

A lion (predator) capures a zebra (prey).

★ **Chameleons** are lizards that use the color of their skin to camouflage themselves. They can quickly change color. This abil-ity allows them to blend in with their sur-roundings, such as rocks or leaves, so predators cannot see them.

★ **Porcupines** have sharp spikes, known as quills, which will hurt a possible predator. When alarmed, a porcupine raises its quills and vibrates them to produce a rat-tling sound. If that does not work, the por-cupine charges backwards with its quills.

Why are sharp quills essential to a porcupine?

ENERGY AND NUTRIENTS IN AN ECOSYSTEM

Ecosystems have producers, consumers, and decomposers:

★ **Producers.** The plants in the ecosystem produce their own food from sun-light through photosynthesis. All the food and energy available in the ecosys-tem is created by plants.

★ **Consumers.** The animals in the ecosystem do not make their own food. They must eat plants or animals. **Herbivores** are animals, like cows, that eat only plants. **Carnivores** are animals, like lions, that eat only animals. **Omni-vores** are animals that eat both plants and animals.

★ **Decomposers.** Some living things in the ecosystem, like ants, worms, fungi, and bacteria, live by breaking down waste products and dead things. These are known as **decomposers.** They put nutrients back into the soil.

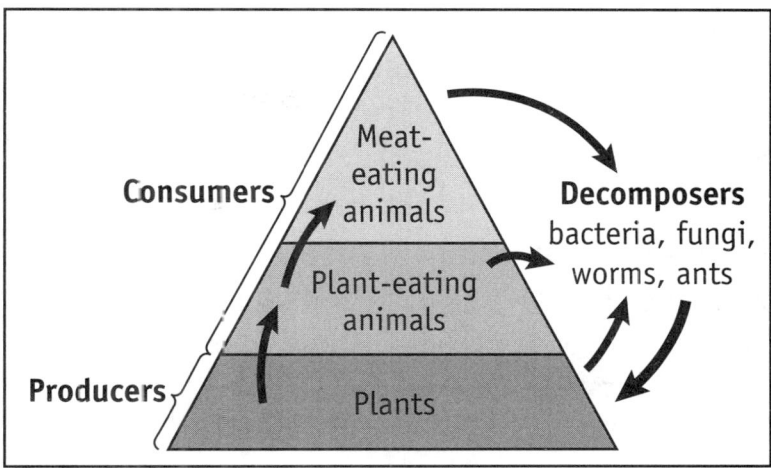

Energy and nutrients are continually recycled in an ecosystem.

APPLYING WHAT YOU HAVE LEARNED

Examine the following list of organisms in a food chain. Show whether they are producers, consumers, or decomposers:

Organism	Producers, Consumers, or Decomposers?	Organism	Producers, Consumers, or Decomposers?
Worms		Bacteria	
Deer		Mice	
Pine trees		Wheat	
Bears		Algae	
Rabbits		Sparrows	
Ants		Frogs	

FOOD CHAIN

A **food chain** shows the relationship between living things in an ecosystem. It shows who eats what. Here is a food chain from a prairie ecosystem:

In this food chain, rabbits eat the grass. Then coyotes eat the rabbits. In this example, the grass is able to store energy from sunlight (*photosynthesis*). The rabbits take this energy when they eat the grass. In this way, a food chain traces the **flow of energy** in an ecosystem. The direction of the arrows shows how the energy moves.

FOOD WEB

A **food web** shows how several living things in an ecosystem interact together. Here is that same prairie system shown as a food web:

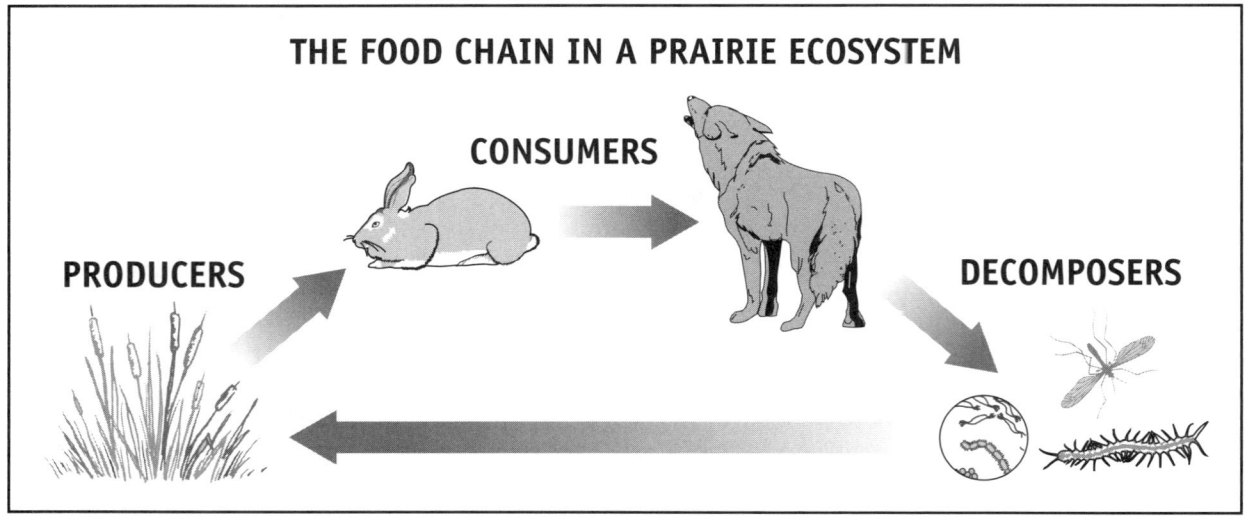

Not only energy, but many **nutrients** are recycled in an ecosystem. For example, plants have special chemicals that all animals need. When the rabbits eat these plants, they absorb these chemicals. When the coyotes eat the rabbits, they take in the same chemicals.

When the prairie grasses, rabbits and coyotes die, their bodies decay. Ants, bacteria and other decomposers break down their remains and return these chemicals to the soil as nutrients. From the soil, these nutrients are absorbed by the roots of plants. Then the cycle begins all over again.

APPLYING WHAT YOU HAVE LEARNED

◆ Why is it important that chemicals in the ecosystem are recycled? _____

_____

_____

_____

_____

Plants and animals must protect themselves from animals that might eat them. In order to survive, each living thing has special adaptive characteristics that make it different from other living things in the same ecosystem. These characteristics give it a special **niche** — a place in that ecosystem based on its unique characteristics.

APPLYING WHAT YOU HAVE LEARNED

Algae
produce
their own
food

Prawns
eat the
algae

Small fish
eat algae
and other
small fish

Sharks
eat the
fish

Seagulls
eat the
fish

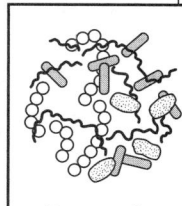

Bacteria
break down
wastes and
decaying
bodies

Look at the organisms living together in an ecosystem near the surface of the ocean. Complete the food web below describing this ecosystem:

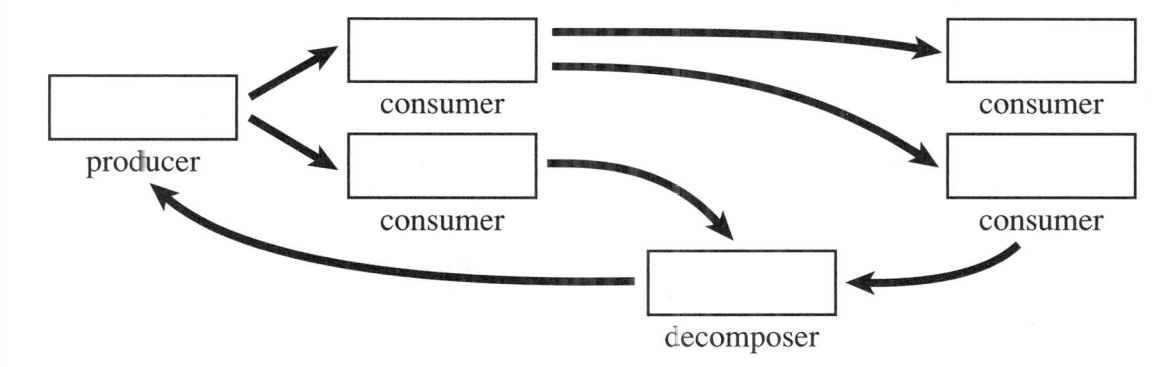

producer

consumer

consumer

consumer

consumer

decomposer

ENVIRONMENTAL CHANGE

Sometimes the physical environment of a place changes. Change can be slow, taking thousands or even millions of years. For example, scientists believe the Sahara Desert was once an ocean. Today, it is a dry desert. This change probably took millions of years. Other changes may happen more quickly. A fire may burn down a forest. The plants and animals in the forest ecosystem may not be able to survive in the new environment.

When environments change slowly, living things are often able to adapt to the changing conditions. When change happens fast, many cannot adapt. Some types of organisms will no longer be able to live in that area. New types of organisms will take their place with adaptive characteristics that help them survive in that environment.

APPLYING WHAT YOU HAVE LEARNED

◆ A farm is abandoned by its owner. Over the next 200 years, it becomes part of the neighboring forest. What kinds of changes would take place there?

_____

_____

_____

_____

_____

WHAT YOU SHOULD KNOW

A. You should know that an **ecosystem** is made up of all the living and nonliving things in a particular area. Each living thing in an ecosystem has its own **habitat** where it can live.

B. You should know that living things in an ecosystem depend on both their physical environment and one another in order to survive.

C. You should know that similar living things in an ecosystem compete with each other for resources, such as oxygen, water, food or space.

D. You should know that living things in an ecosystem have **adaptive characteristics** which help them to survive and reproduce. These special characteristics give them a **unique niche**, or role, in the ecosystem.

E. You should know that living things in an ecosystem often modify their physical environment.

F. You should know that changes in an environment can cause some types of living things to die out.

CHAPTER STUDY CARDS

Ecosystems

Ecosystem. All the living and nonliving things in an area.

★ **Adaptive Characteristics.** The characteristics of a plant or animal that help it survive and reproduce in its environment.

★ **Competition.** Plants and animals often compete with each other for the same resources.

★ Animals sometimes **modify** their environment to meet their needs.

Flow of Energy in an Ecosystem

★ **Producers.** Plants produce their own food.

★ **Consumers.** Animals eat plants or animals for energy. **Predators** are animals that hunt and kill **prey**.

★ **Decomposers.** Ants, bacteria, and fungi break down wastes and dead plants and animals; return nutrients to the soil.

★ **Food Chain/ Food Web.** Diagrams that show how energy (food) flows through an ecosystem; who eats what.

CHECKING YOUR UNDERSTANDING

1 Which correctly shows a food chain in the ecosystem to the right?

A Grass → cow → human

B Caterpillar → tree → human

C Cow → grass → human

D Tree → bird → caterpillar

OBJ. 2
2.9 (B)

To answer this question correctly, you must understand that a food chain shows the flow of energy in an ecosystem. Food is created by plants using the energy of the sun (photosynthesis). These plants in turn are eaten by animals. Other larger and faster animals may eat these animals. In this case, the correct order would start with grass. Cows are animals that eat grass. Humans eat beef from cows. Therefore, the correct answer would be **Choice A**.

Now try answering some additional questions on your own:

2   An antelope is an animal that eats grass on the plains of Africa. Antelopes are the prey of lions, tigers, and other predators. Which of the following characteristics would be an advantage for an antelope?

F   Brightly colored fur   **H**   Sharp claws   OBJ. 2
G   Fast, strong legs   **J**   Small eyes   5.9 (A)

3   Lions hunt antelopes and other animals that eat grass and green plants. Which of the following characteristics would NOT be an advantage to a lion?

A   A long tail   **C**   Fast legs   OBJ. 2
B   Sharp teeth   **D**   Large size   5.9 (B)

4   Decomposers are helpful to an ecosystem because they —

F   return nutrients to the soil   OBJ. 2
G   eat producers   5.5 (B)
H   make food with photosynthesis
J   are food for consumers

5   In what order do the owl, acorn, and squirrel form a food chain?

A

B

OBJ. 2
5.5 (B)

C

D

6   If foxes were removed from this food web, which animal population would most likely increase?

F   Snakes   OBJ. 2
G   Rabbits   2.9 (B)
H   Birds
J   Insects

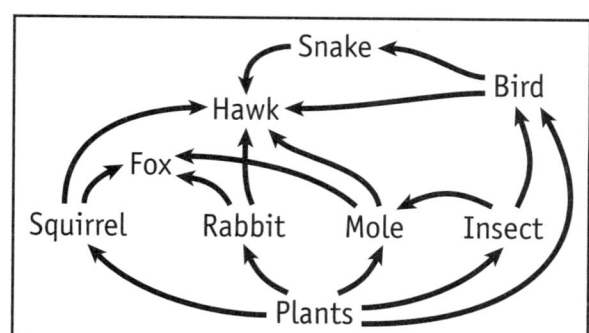

7   What basic needs do bees meet by building a bee hive?

A   Food   **C**   Air   OBJ. 2
B   Water   **D**   Shelter   3.8 (D)

8   **Which group shows the correct order of a prairie food chain?**

   F   grasshopper $\longrightarrow$ grass $\longrightarrow$ snake $\longrightarrow$ toad $\longrightarrow$ hawk

   G   grass $\longrightarrow$ grasshopper $\longrightarrow$ snake $\longrightarrow$ toad $\longrightarrow$ hawk

   H   grass $\longrightarrow$ grasshopper $\longrightarrow$ toad $\longrightarrow$ snake $\longrightarrow$ hawk

   J   hawk $\longrightarrow$ grasshopper $\longrightarrow$ toad $\longrightarrow$ grass $\longrightarrow$ snake

OBJ. 2
5.5 (B)

Use the picture below and what you know about plants and animals to answer questions 9 – 10.

   Tropical rain forests have high temperatures and heavy rainfall. They are home to many different types of plants and animals.

9   **Which part of a bird in the rainforest most helps it escape predators?**

   A   Long, thin legs

   B   Strong wings

   C   Small head

   D   Brightly colored feathers

OBJ. 2
5.9 (A)

10   **In the rainforest, herbivores would most likely compete for —**

   F   oxygen

   G   sunlight

   H   water

   J   food

♦ Examine the Question
♦ Recall What You Know
♦ Apply What You Know

OBJ. 2
3.8 (B)

11 **Which information can be found on a food chain?**

   A The habitat needs of predators

   B How climatic conditions affect organisms

   C The relationship between plants and sunlight

   D The relationship between prey and predatory animals

OBJ. 2
5.5 (B)

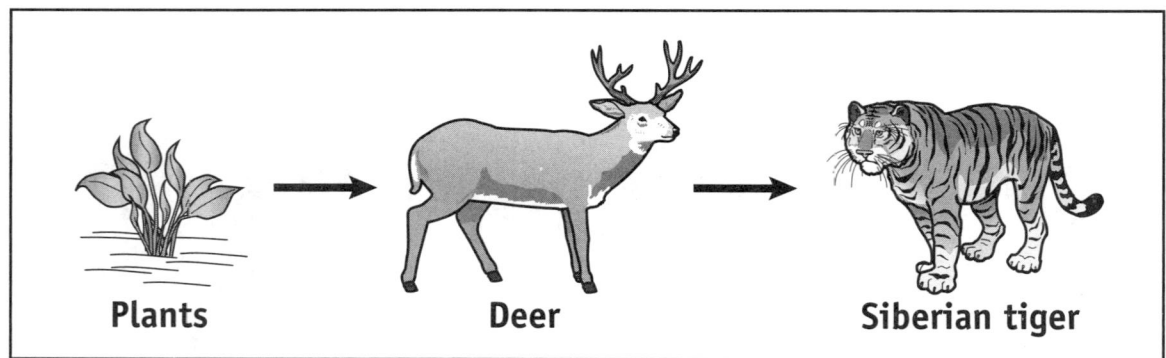

| Plants | Deer | Siberian tiger |

12 **What would be the best title for the diagram above?**

   F Plant and animal differences

   G Animals that are threatened

   H All the living things in an ecosystem

   J Energy flow in an ecosystem

OBJ. 2
5.5 (B)

13 **In one study, scientists found that the number of deer increased when humans built houses in their habitats. Which best explains why the number of deer increased?**

   A Deer no longer lived in wild areas.

   B Deer found less food living near humans.

   C More predators of deer live near people's homes.

   D People destroyed deer predators when they built homes.

OBJ. 2
3.8 (C)

FOOD CHAIN

Green plants ⟶ **?** ⟶ Frogs ⟶ Snakes

14 **Which of the following would best complete this food chain?**

   F Lemons

   G Tigers

   H Insects

   J Owls

♦ Examine the Question
♦ Recall What You Know
♦ Apply What You Know

OBJ. 2
5.9 (B)

CHAPTER 6

INHERITED TRAITS AND LEARNED BEHAVIOR

In this chapter, you will learn the difference between learned behavior and inherited traits.

— MAJOR IDEAS —

★ Some traits are inherited by plants and animals from their parents.
★ Differences in inherited traits affect the ability of animals and plants to survive changes in the environment.
★ Animals are able to learn some behaviors as they interact with their environment.

INHERITED TRAITS

All plants and animals **inherit** characteristics from their parents. These inherited traits are passed on from the parents to their children. For example, a dog inherits the color of its fur from its parents. When you inherit a trait, like eye color or height, you cannot change it.

APPLYING WHAT YOU HAVE LEARNED

✦ As you just read, children inherit many of their characteristics from their parents. Describe some of the traits you inherited from your parents.

_____

_____

_____

APPLYING WHAT YOU HAVE LEARNED

Some people show dimples in their cheeks when they smile. Others, no matter how hard they try, cannot do this. When people put their hands together and interlock their fingers, they place either their left or right thumb on top. Other people can curl the sides of their tongue. Some people are able to make a "Vulcan" hand sign like Mr. Spock did on Star Trek. These are all inherited traits. A person who cannot curl his or her tongue cannot learn to do so. In addition, some people have a regular thumb, while others have a "Hitchhiker's thumb" which can be bent farther backwards.

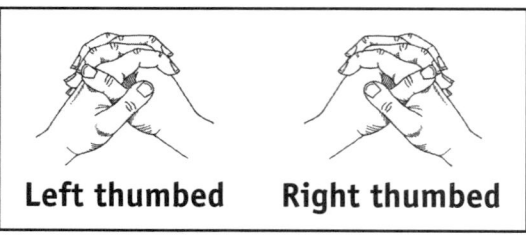

Left thumbed   Right thumbed

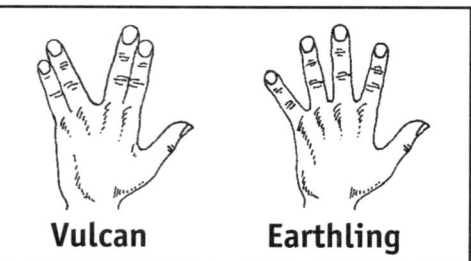

Vulcan   Earthling

Hitchhiker's thumb   Regular thumb

Pick a classmate and make a list of these easily observable traits. Compare your traits with those of your partner:

Characteristic	Your Trait	Classmate's Trait
1. Dimples when smiling	☐	☐
2. Right thumb on top	☐	☐
3. Able to curl tongue	☐	☐
4. "Vulcan" hand sign	☐	☐
5. "Hitchhiker's" Thumb	☐	☐

INSTINCTIVE BEHAVIOR

Not all traits are physical characteristics, like hair color or size. Some traits are behaviors — the way animals and, in some cases plants, act. For example, you blink your eyes and swallow food when you eat. You did not have to learn this behavior. You already knew how to do it when you were born.

Animals have many types of behaviors they inherit. For example, birds know how to fly. They also know how to fly in certain patterns in groups. These inherited behaviors are called **instinctive behavior**.

LEARNED BEHAVIOR

Animals often respond to their environment. For example, if a wild deer sees a new type of flower, it might eat it. If the flower tastes good, the deer will eat it again. This is called **learned behavior**. An animal does not inherit learned behavior. Learned behavior happens when an animal learns to do or not do something based on what happened to it when it did that action before.

APPLYING WHAT YOU HAVE LEARNED

★ Describe something you have noticed about yourself. For example, you speak loudly, or you have curly hair. Is this a learned behavior or an inherited trait? _____ Explain _____

_____

_____

_____

★ Classify the characteristics below as inherited traits or learned behavior:

Characteristic	Inherited	Learned
1. Can roll tongue	☐	☐
2. Eye color	☐	☐
3. Likes to read books	☐	☐
4. Likes to drink soda	☐	☐

SURVIVING ENVIRONMENTAL CHANGE

Inherited traits, like other adaptive characteristics, can help living things survive in their environment. If the environment changes, some inherited traits may become more helpful or advantageous.

For example, some rabbits are white. The whiteness of a rabbit's fur is an inherited trait. If the climate of an area is cold and there is snow, the whiteness of a rabbit's fur may become more of an advantage. Predators like foxes are less able to see white rabbits in the snow. With increased snow, more white rabbits will survive than darker rabbits. Because more of them will survive, they will have more children. Gradually, the fur of the rabbits living in the area will become whiter.

Why are dark rabbits at a disadvantage in snow?

Learned behaviors can also help animals to survive environmental change. Animals can learn new ways to adapt to their environment. For example, deer that once ate wild plants may learn to eat fruits or flowers growing in people's gardens. However, learned behaviors are not inherited. Each new group of animals has to learn the behavior from their parents or discover it on their own.

WHAT YOU SHOULD KNOW

A. You should know that some traits are inherited by plants and animals from their parents.

B. You should know that differences in inherited traits affect the ability of animals and plants to survive changes in the environment.

C. You should know that animals are able to learn some behaviors as they interact with their environment.

CHAPTER STUDY CARDS

Inherited Traits

An **inherited trait** is something inherited by a plant or animal from its parents.

★ Many inherited traits are physical characteristics — like hair color or eye color.

★ **Instinctive Behaviors**. Some inherited traits are instinctive behaviors — such as birds flying south in the winter or bears hibernating.

★ Inherited traits cannot be changed.

Learned Behavior

Learned behavior is something an animal learns by interacting with the environment.

★ Learned behavior is **not** inherited.

★ Learned behavior can be changed.

★ Learned behavior occurs when an animal has an experience it seeks to avoid or repeat. For example, a coyote learns to avoid cacti.

CHECKING YOUR UNDERSTANDING

The illustration below shows a type of bat as it is flying in the dark making a high-pitched sound.

1 **Which behavior of the bat is most likely a learned behavior?**

OBJ. 2
5.10 (B)

   A Flying
   B Sleeping upside down
   C Using sound to find objects
   D Avoiding humans

Sound waves

This question tests if you know the difference between learned behaviors and inherited traits. Some behaviors are inherited or instinctive. An animal just knows how to do them. Other animal behaviors are learned after interacting with the environment. For example, all bats fly, sleep upside down and use sound when flying. Not all bats know to avoid humans. This has to be learned. Therefore, **Choice D** is the best answer.

Now try answering some additional questions on your own:

2 **Margaret has just received a puppy for her tenth birthday. In what ways must her new puppy be like its parents?**

OBJ. 2
5.10 (A)

   F Will always bark when it wants to go for a walk
   G Will always learn the same tricks
   H Will have a similar body fur color
   J Will stay living in the same place

3 **Which of the following is NOT an inherited trait of coyotes?**

   A Eye color
   B Favorite food
   C Fur color
   D Size of teeth

   ◆ Examine the Question
   ◆ Recall What You Know
   ◆ Apply What You Know

OBJ. 2
5.10 (A)

Use the chart below to answer the next question:

Student's Name	Student's Eye Color	Student's Height
Dwayne	Brown	137 cm
Marcus	Brown	139 cm
Stephanie	Blue	132 cm
Carrie	Green	137 cm

4 Which students would most likely have parents with brown eyes?

F   Stephanie and Marcus

G   Dwayne and Stephanie

H   Carrie and Stephanie

J   Dwayne and Marcus

♦ **Examine the Question**
♦ **Recall What You Know**
♦ **Apply What You Know**

OBJ. 2
5.10 (A)

5 Young spiders spin webs to catch insects to eat. How do young spiders know how to build such webs?

A   They practice each day.

B   They build webs in response to their environment.

C   They inherit this behavior from their parents.

D   They learn from watching other spiders.

OBJ. 2
5.10 (B)

6 Kittens have some characteristics that are inherited and others that are learned from their environment. Which of the following traits of a kitten is learned?

F   The color of its fur

G   The size of its ears

H   Playing with a ball

J   Making a purring sound

OBJ. 2
5.10 (B)

The picture to the right shows several types of birds.

7 What inherited trait do all these birds share?

A   They can fly.

C   They have beaks.

B   They have webbed feet.

D   They eat worms.

OBJ. 2
5.10 (A)

8 **Which of these is an example of a learned behavior?**   OBJ. 2
5.10 (B)

F

H

G

J

Examine the drawing below

9 **The butterfly above has spots on its wings. The spots on the butterfly's wings are —**

A important for finding food
B an inherited trait
C a learned behavior
D an advantage in a cold climate

OBJ. 2
5.10 (A)

CHECKLIST OF OBJECTIVES IN THIS UNIT

Directions. Now that you have completed this unit, place a check (✔) next to those objectives you understand. If you are having trouble recalling information about any of these objectives, review the chapter shown in the accompanying brackets.

- [ ] You should be able to identify traits that are inherited in plants and animals from their parents. **[Chapter 6]**
- [ ] You should be able to give examples of learned behavior that result from the influence of the environment. **[Chapter 6]**
- [ ] You should be able to compare the adaptive characteristics of different groups of plants and animals that improve their ability to survive and reproduce in an ecosystem. **[Chapter 5]**
- [ ] You should be able to analyze and describe adaptive characteristics that result in a niche in an ecosystem for each type of living thing. **[Chapter 5]**
- [ ] You should be able to predict some adaptive characteristics required for survival and reproduction by living things in an ecosystem. **[Chapter 5]**
- [ ] You should be able to describe and compare life cycles of plants and animals. **[Chapter 4]**
- [ ] You should be able to observe and describe the habitats of different forms of life within an ecosystem. **[Chapter 5]**
- [ ] You should be able to observe and identify forms of life with similar needs that compete with one another for resources such as oxygen, water, food, or space. **[Chapter 5]**
- [ ] You should be able to describe environmental changes in which some forms of life would do well, become ill, or die out. **[Chapters 5 and 6]**
- [ ] You should be able to describe how living organisms modify their physical environment to meet their needs, such as beavers building a dam or humans building a home. **[Chapter 5]**
- [ ] You should be able to identify the characteristics of different kinds of plants and animals that allow their needs to be met. **[Chapter 4]**
- [ ] You should be able to compare and give examples of the ways living things depend on each other and on their environments. **[Chapters 5]**
- [ ] You should be able to identify patterns of change such as in metamorphosis. **[Chapters 4 and 5]**

THE PHYSICAL SCIENCES

UNIT 3

In this unit, you will learn about the physical sciences. The **physical sciences** study matter and energy. **Matter** is everything around you. It includes anything that takes up space and has **mass** — almost everything you can see.

If matter can be thought of as everything around you,

Students experimenting with different forms of matter

energy is what moves matter. In this unit, you will learn that energy has the ability to move or change matter. Electricity, heat and light are all different forms of energy.

★ **Chapter 7: The Properties of Matter**

In this chapter, you will learn about matter and its properties, such as mass and conductivity. You will also learn how matter can change its form, from solid to liquid to gas. Finally, you will learn how different types of matter form mixtures.

★ **Chapter 8: Motion, Force and Energy**

This chapter looks at how matter moves and how force is needed for matter to move or change. You will also learn how energy provides this force. There are many types of energy, and one type of energy can be changed into another.

CHAPTER 7

THE PROPERTIES OF MATTER

In this chapter, you will learn about the properties of matter.

— MAJOR IDEAS —

★ **Matter** can be classified based on its properties. Some of the properties of matter are mass, magnetism, and the ability to conduct heat, electricity or sound.

★ **Matter** has three states: *solid*, *liquid*, or *gas*

★ Each **substance** has some characteristics that stay the same, like its **boiling** and **melting** points.

★ **Mixtures** combine different kinds of matter (**substances**).

★ When a **solution** is formed, some of the physical properties of its ingredients may change.

WHAT IS MATTER?

What do a strawberry, a chocolate cake and air have in common? If you said they were different types of matter, you would be correct.

Matter is the stuff of the universe. Matter is everything that takes up space (*volume*) and has mass. Matter comes in many different shapes and sizes. For example, air and water are both forms of matter. The book you are looking at now is matter. Plants and animals are matter. Diamond rings, steam from a boiling tea kettle, sand on a beach, and the clouds in the sky are all different forms of matter.

However, **NOT EVERYTHING** is matter. Light and electricity are not matter because they do not have mass and they do not take up their own separate space.

APPLYING WHAT YOU HAVE LEARNED

Identify examples of matter and things that are not matter.

Matter	Not Matter
1. _____	1. _____
2. _____	2. _____
3. _____	3. _____

MASS AND WEIGHT

Mass is *how much* there is of an object — the *amount* of an object. Scientists usually measure mass in **grams** (g) or **kilograms** (kg). You may remember that scientists use a **double-pan** or **triple-beam balance** to measure mass.

Although mass and weight are related, they are different things. On Earth, an object's **mass** tells us how much there is of an object. Its **weight** is the force of attraction created by **gravity**. Our weight would change if we went to the moon or some other planet, but not our mass. The chart below shows what you would weigh on Mars or the moon if you weighed 80 pounds on Earth. However, your **mass** on Mars or the moon *would stay the same*.

A Person's Mass	Earth Weight	Mars Weight	Moon Weight
	80 pounds	30 pounds	13 pounds

THE PROPERTIES OF MATTER

People use a variety of different ways to describe an object — its shape, its color or texture. Scientists also use different ways to describe matter. Every type of matter has certain **properties**. A **property** is a **characteristic** or quality that describes a particular type of matter. **Mass** is one property scientists use to describe a particular piece of matter. Other properties of matter include the following:

Is it solid, liquid or gas?	Does it conduct heat, sound or electricity?	Is it magnetic?

Scientists are able to use these properties to describe and classify different kinds of matter. Let's examine each of these properties more closely.

THE THREE STATES OF MATTER

Each type of matter can exist in three **states** or forms — as a **solid**, **liquid** or **gas**. Although you might think of iron as a solid, when heated to a very high temperature, it will melt. You may think of air as a gas, but if it cools down enough, it will become liquid. The reason for these different states is that all matter is actually made up of tiny particles. These particles are so small we can't see them. Scientists believe these particles are constantly moving.

SOLIDS

In a **solid**, these tiny particles are locked into fixed positions. Since they are locked into position, they give the substance a fixed **volume** and **shape**. Volume, you may remember, is how much space something takes up. The particles of a solid are moving, but they vibrate in place. A rock, an ice cube and a diamond are all examples of a solid.

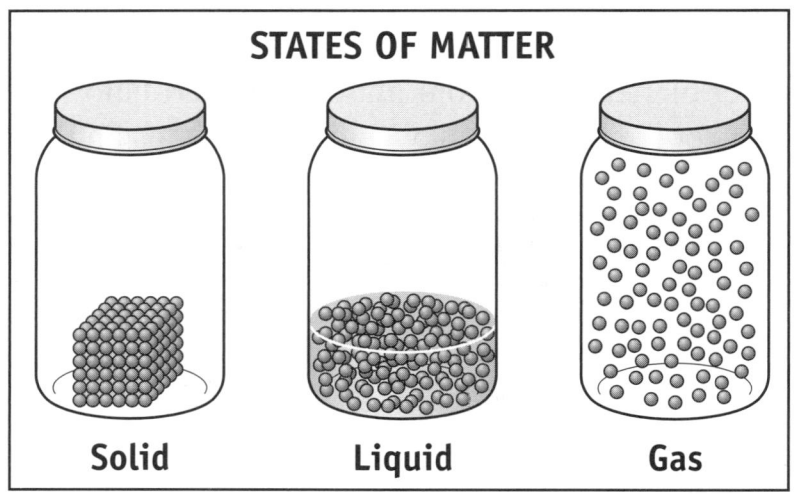

STATES OF MATTER

Solid   Liquid   Gas

LIQUIDS

When heat is applied to a solid, its particles begin to vibrate faster. Eventually, its particles vibrate so much they start to move around each other. The solid melts and becomes a liquid. Some examples of liquids at room temperature are water, milk, and mercury. Since the particles of a liquid can move around each other, a liquid can change its shape easily. A liquid will take the shape of whatever container it is in. For example, if you pour milk from a large container into a smaller glass, the milk will take the shape of the glass it is poured into. Its volume stays the same.

GASES

If heat is applied to a liquid, its particles move around even faster. Eventually, its tiny particles move so rapidly they spread out in all directions as a **gas**. A gas has no fixed shape and no fixed volume. It will fill up whatever space it has. Some examples of common gases are oxygen and carbon dioxide.

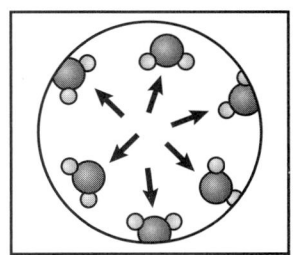

Water has a special characteristic. Even if it does not boil, an open container of water will **evaporate** into the air. When water **evaporates**, it turns from a liquid into a gas.

APPLYING WHAT YOU HAVE LEARNED

How do solids, liquids and gases differ from each other? Complete the following chart by filling in the characteristics called for.

Characteristic	Solid	Liquid	Gas
Speed of Particles	Particles are rigid and do not move		
Shape			Has no fixed shape
Volume		Has a fixed volume	

MELTING AND BOILING POINTS

Do you know the difference between melting and boiling points?

★ **Melting Point.** The temperature when a substance turns from a solid to a liquid.

★ **Boiling Point.** The temperature when a liquid turns into a gas.

A **substance** will always have the same melting and boiling point. For example, water always freezes at 0°C and boils at 100°C. That means that ice will melt once it rises above 0°C. Water will boil once it reaches 100°C. Because substances have fixed melting and boiling points, you always see them in the same state at room temperature — iron is solid, water is liquid and oxygen is gas.

APPLYING WHAT YOU HAVE LEARNED

✦ Describe what will happen to the water in each beaker:

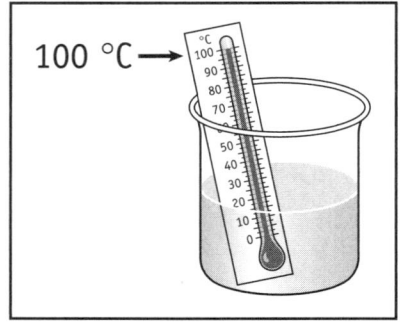

100 °C →

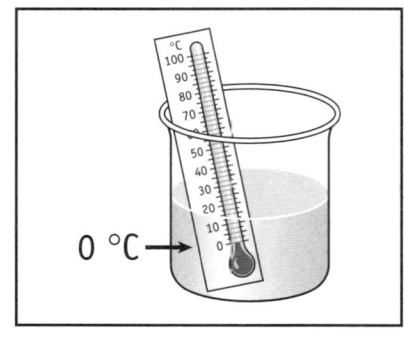

0 °C →

_____   _____

_____   _____

✦ The diagram to the right shows a beaker with water in it. The dots indicate particles of water. Describe what is taking place in the beaker.

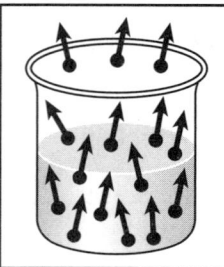

_____

_____

MAGNETISM

A **magnet** will attract some metals, such as **iron**, **nickel**, **cobalt** and **steel**. A magnet will pull pieces of those metals towards it, or even pick them up. Those metals are **magnetic**. However, many types of metals are not attracted to magnets. A magnet will have no effect at all on tin, aluminum, copper, gold or silver.

A magnet also has no effect on non-metals, like plastic, rubber or wood. Scientists can use magnetism to separate magnetic objects from non-magnetic ones.

APPLYING WHAT YOU HAVE LEARNED

A small sliver of iron accidently flew into a child's eye. How could doctors use their knowledge of magnetism to help the child? _____

_____

APPLYING WHAT YOU HAVE LEARNED

Which of the following items is magnetic?

☐ A copper penny   ☐ A nickel chain   ☐ A plastic hanger
☐ Aluminum scissors   ☐ A steel toy car   ☐ A rubber band

CONDUCTIVITY

Another important property of matter is how well it carries heat, sound or electricity. This ability is known as **conductivity**.

HEAT

Some materials carry heat better than others. In objects made from these materials, heat moves faster from one end to the other. For example, metals like copper and aluminum conduct heat well. Wood, plastic and rubber do not conduct heat very well. For this reason, wood, plastic and rubber are often used for pot or pan handles. They will not get hot.

SOUND

Some materials conduct sound better than others. For example, metal or wood will conduct sound better than rubber. Speakers on a radio will never be made of rubber.

APPLYING WHAT YOU HAVE LEARNED

Try this experiment tonight at home. Put your ear against a long piece of wood. Ask a parent or friend to tap the other side of the wood. Now cover that end of the wood with a kitchen sponge and have your parent or friend tap on it again.

Which time was the tapping louder? _____ What might explain this? _____

_____

_____

Try this experiment using other materials around your house. Which items conducted sound the best? _____

_____

_____

ELECTRICITY

Some forms of matter are good conductors of electricity. Metals conduct electricity well. Other substances, like carbon and sulfur, do not conduct electricity.

Something that does not conduct heat, sound or electricity is called an **insulator**. It may be used to block the flow of energy. For example, wires are usually made of copper or some other metal that is a good conductor of electricity. However, the wire will always be covered in plastic or some other type of

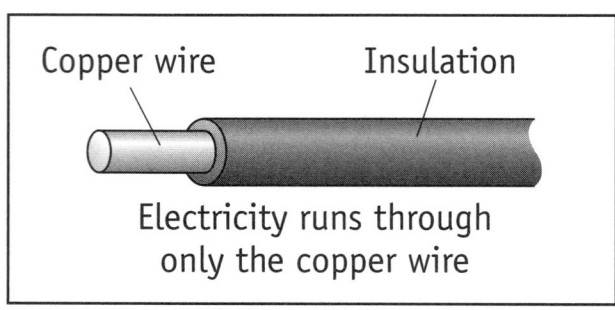

Copper wire   Insulation

Electricity runs through only the copper wire

insulation — a material that does **not** conduct electricity well. In the picture above, the copper wire at the center of the wire is a good conductor of electricity. The copper wire is covered with a plastic material that does not conduct electricity. Someone who touches this insulation will be protected from a shock caused by the electricity running through the wire.

APPLYING WHAT YOU HAVE LEARNED

Answer the following questions.

◆ Why is it important for engineers and architects to know which materials are good conductors and which materials are good insulators? _____

_____

_____

_____

◆ Look around your classroom. Identify four objects that you think are good conductors of heat, electricity or sound.

• _____

• _____

• _____

• _____

MIXTURES AND SOLUTIONS

Do you know what a mixture is? In science, a **mixture** is when different things are mixed together but not chemically combined. Think of a salad. A salad may have lettuce, carrots, tomatoes, cucumbers and other vegetables. All together, these ingredients make up the salad. However, each ingredient in that salad keeps its own characteristics.

CHEMICAL COMBINATION OR MIXTURE?

All matter is made of small particles called atoms. When the atoms combine or join together, a new substance is formed. For example, **water** is actually made of atoms from two gases — oxygen and hydrogen. Water's characteristics are totally different from either gas, because their atoms have combined. In a mixture, the atoms do not combine. Thus, the ingredients in the mixture keep many of their characteristics.

APPLYING WHAT YOU HAVE LEARNED

Identify which of the following ingredients form a mixture and which ingredients are chemically combined:

Ingredients	Mixture	Chemically Combined
✦ Salt and sugar	☐	☐
✦ Carbon dioxide gas (oxygen and carbon)	☐	☐
✦ Water (hydrogen and oxygen)	☐	☐
✦ Salt (chlorine gas and niacin)	☐	☐
✦ Maple syrup (water and sugars)	☐	☐
✦ Air (oxygen, nitrogen and other gases)	☐	☐

There are many kinds of mixtures.

★ **Solid with a Solid.** You can mix two solids together. For example, you could mix sugar or sand (*solids*) with iron filings (*solid*).

★ **Solid with a Liquid.** You could mix a solid with a liquid. For example, you could stir chocolate powder (*solid*) into milk (*liquid*).

★ **Liquid and Gas Mixtures.** Two liquids can be mixed. For example, you can mix water and salad oil. Even gases can be mixed. Air is a mixture of nitrogen, oxygen and other gases. Soda is a mixture of liquid and gas.

APPLYING WHAT YOU HAVE LEARNED

Provide examples for the following mixtures:

◆ **Liquid** mixed with a **solid**: _____

_____

◆ **Solid** mixed with a **solid**: _____

_____

In addition to knowing that there are many kinds of mixtures, you should be aware of three main ideas about mixtures:

★ **In Mixtures, Ingredients Are Not Chemically Joined.** Mixtures are made of ingredients that are mixed together but do not join together chemically. Their atoms stay separated.

★ **Ingredients Keep Many of their Characteristics.** Each ingredient in a mixture will usually keep many of its own characteristics.

★ **Can Be Separated Out.** The ingredients in a mixture can be separated out again by physical methods.

In a mixture of salt and iron, the iron can be removed with a magnet.

Because the ingredients in a mixture are not chemically combined, it is possible to separate them. What is the best way to separate the ingredients in a mixture? That will depend on the properties of the ingredients in the mixture.

There are three common ways of separating ingredients in a mixture:

Separation Methods	How It Can Be Done:
Magnetism	If one of the ingredients is magnetic, you can use a magnet to separate the materials. For example, you could use a magnet to pull out iron filings from sand in a mixture.
Filter	If the ingredients have particles that are different sizes, or one ingredient is a liquid and the other is a solid, you can pour the mixture over a fabric filter or screen. One ingredient will pass through the filter but not the other. For example, to separate sand from ocean saltwater, pour it through a filter.
Boiling or Evaporating	If the mixture has a solid dissolved in water, you can wait for the water to evaporate or boil the water. The water will disappear, but the solid will remain at the bottom of the container. For example, if salt is dissolved in water, you can boil the solution. The water will evaporate, but the salt will remain.

APPLYING WHAT YOU HAVE LEARNED

✦ A student places sugar and ground-up cinnamon in a cup and shakes it. What will the new mixture be like? Do the ingredients keep their characteristics? _____

_____

_____

_____

✦ A student mixes oil and water and stirs them. Later, the student returns and sees that the oil has formed a layer on the top and the water is on the bottom. Has the student created a mixture? Explain your answer. _____

_____

_____

_____

SOLUTIONS

A **solution** is a special kind of mixture. In a solution, one substance dissolves into another. Think of what happens when a spoonful of sugar is stirred in a glass of hot tea. The sugar seems to disappear into the tea.

MIXTURE OF TEA AND SUGAR

Sugar dissolves in tea to form a solution.

Where does the sugar go? The sugar actually breaks down into tiny particles. These sugar particles are surrounded by water particles (*molecules*). That is why the sugar seems to disappear.

Not every mixture is a solution. Only some things will dissolve in others. When something will not dissolve, you can still see it. For example, if you stir sand in a container of water, you would have a cloudy mixture — but not a solution. After awhile, the sand will sink to the bottom of the container and the water will stay on top.

APPLYING WHAT YOU HAVE LEARNED

Which of the following form solutions and which do not?

Ingredients	Solution	Not a Solution
✦ Oil and water	☐	☐
✦ Milk and water	☐	☐
✦ Salt and water	☐	☐
✦ Pepper and water	☐	☐

In a solution, the ingredients may change some of their properties and keep others. The sugar that dissolved in the tea no longer looks like white crystals, but the tea / sugar solution tastes sweet from the sugar.

SEPARATING INGREDIENTS FROM A SOLUTION

Just like other mixtures, the ingredients in a solution can be separated again. The most common way to separate the ingredients of a solution is to boil the solution.

When you boil a solution, the liquid ingredient turns to a gas. The solid ingredient will remain in the container. The amount of solid left will equal the amount of solid that was put into the solution to begin with.

Miranda mixes 50 grams of salt in 0.5 liters of water. She stirs it completely until she can no longer see the salt.

To separate the water and the salt, Miranda boils the solution. She continues to boil it until all the water has evaporated.

Miranda sees salt at the bottom of the container. What is the mass of that salt?

_____

_____

_____

Here is another way to separate parts of a solution. If you put a stick in a sugar-water solution, sugar particles will collide with the stick and stick to it. Eventually, the particles will form sugar crystals around the stick.

WHAT YOU SHOULD KNOW

A. You should know that matter can be classified based on its properties. Some of the properties of matter include its mass, magnetism, and the ability to conduct heat, electricity or sound.

B. You should know that matter has three states: **solid**, **liquid**, or **gas**.

C. You should know that each **substance** has some characteristics that stay the same, like its **boiling** and **melting** points.

D. You should know that **mixtures** are made of different kinds of matter (**substances**). When a **solution** is formed, the physical properties of its ingredients change.

CHAPTER STUDY CARDS

Properties of Matter

Every type of matter has certain **properties**.

★ **Mass.** How much there is of an object — the amount of matter is measured in grams (g) and kilograms (kg).

★ **States of Matter.** The three states of matter are *solid*, *liquid*, and *gas*.

★ **Magnetism.** A force of attraction between a magnet and certain types of metals.

★ **Conductivity.** How well something carries heat, sound, or electricity.

States of Matter

Matter can exist in one of three forms:

★ **Solid.** It has a **fixed** **volume** and **shape**.

★ **Liquid.** A liquid takes the shape of whatever container it is in; its volume stays fixed.

★ **Gas.** Its particles move about in all directions. It has no fixed shape and no fixed volume.

Melting and Boiling Points

★ A substance will always have the same melting and boiling point.
  • Water freezes at 0° C
  • Water boils at 100° C

★ When ice melts, it changes its **state** from a **solid** to a **liquid.**

★ When water boils, it changes its state from a **liquid** to a **gas.**

Mixtures and Solutions

★ **Mixture.** When different materials are combined, they are called a **mixture.** The ingredients in the mixture can be separated out by physical means — such as a filter, a magnet or boiling.

★ **Solution.** A solution is a special kind of mixture. The ingredients are broken down and completely mixed. When a liquid solution evaporates, the solid ingredient in the solution is left.

CHECKING YOUR UNDERSTANDING

1 Which illustration below shows a liquid being changed into a gas? OBJ. 3 5.7 (A)

 **A**
 **B**
 **C**
 **D**

HINT

To answer this question, you need to understand the different states of matter — solid, liquid, and gas. Picture A shows a solid being cooked; Picture B shows a liquid being boiled; Picture C shows popcorn being popped over a flame. Picture D shows a liquid (pancake batter) being poured onto a hot skillet. Therefore, **Picture B** is the correct choice. When a liquid boils, it changes into a gas.

Now try answering some additional questions on your own:

2   **Sugar and water are poured into a container and mixed together. Which method would be the fastest way to separate the sugar from the water-sugar solution?**

    F   Place a stick in the solution so that sugar particles cling to it.
    G   Burn up the solution.
    H   Put a magnet into the solution.
    J   Pour the solution through a fabric filter.

OBJ. 3
5.7 (C)

3   **A student dissolves salt in water. Which best describes what happens to the water?**

    A   It disappears into the salt.
    B   It changes its taste.
    C   It evaporates.
    D   It changes its color.

◆ **Examine the Question**
◆ **Recall What You Know**
◆ **Apply What You Know**

OBJ. 3
5.7 (C)

4   **Which of the following objects would be attracted to a magnet?**

    F   Paper bag   **H**   Copper penny
    G   Rubber ball   **J**   Iron nail

OBJ. 3
5.7 (A)

5   **Matter can change its form. What happens when water freezes?**

    A   Liquid becomes a gas
    B   Gas becomes a liquid
    C   Liquid becomes a solid
    D   Solid becomes a liquid

OBJ. 3
5.7 (A)

6   **Which of these would be attracted to a magnet?**

    F   Gold   **H**   Wood
    G   Glass   **J**   Iron

OBJ. 3
5.7 (A)

The picture below shows a nail being attracted by object X

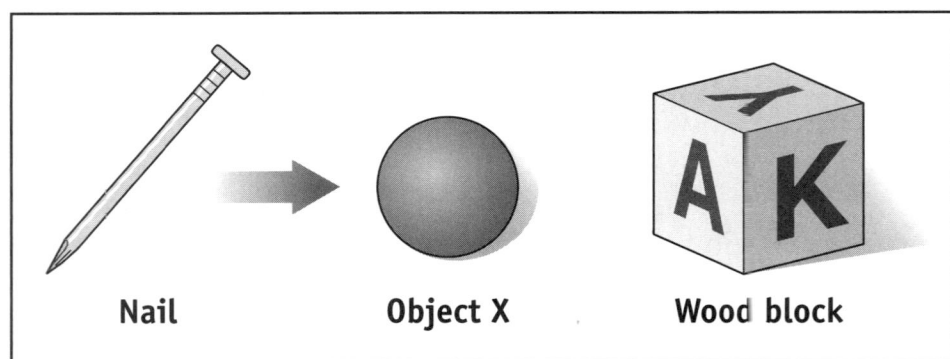

Nail   Object X   Wood block

7   The nail is pulled by object X, but the wood block next to it does not move. What property does object X have that attracts the nail?

A   Density

B   Magnetism

C   High temperature

D   Solid state

◆ Examine the Question
◆ Recall What You Know
◆ Apply What You Know

OBJ. 3
5.7 (A)

8   Iron filings are a black powder. When iron filings are mixed with sand, they create a greyish powder. Once the ingredients have been mixed together, what would be the best way to separate the ingredients of this mixture?

F   Put the mixture in water and stir it

G   Use a magnet to separate the iron filings

H   Pour the mixture through a filter

J   Bring the mixture to a boil

OBJ. 3
5.7 (B)

9   An ice cream bar is left out in the sun. All of the following are likely to happen to the ice cream EXCEPT —

A   melt

B   lose its shape

C   become a liquid

D   become a gas

OBJ. 3
5.7 (A)

10   Which of the following is an example of an object changing from a liquid to a gas?

F   A glass dish breaks.

G   A tire loses air.

H   A pot of soup boils.

J   A piece of plastic catches on fire.

OBJ. 3
5.7 (A)

11 The illustration to the right shows an open jar of water with the sun shining on it. Justin put the jar of water on a picnic table outside in the sunlight. Which of the following pictures shows what probably happened after several hours? OBJ. 3 5.7 (A)

  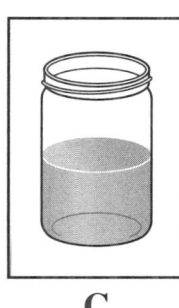 

A   **B**   **C**   **D**

12 A student placed a metal spoon and a wooden spoon into a glass of very hot water. He waited several minutes and then touched the spoon handles. What most likely happened?

F The wooden handle became hot.
G Both handles became hot.
H Neither handle became hot.
J The metal handle became hot.

OBJ. 3 5.7 (A)

13 A student dissolves 50 grams of salt in 1 liter of water. The water is heated until it boils. After the water evaporates, how much salt will be left?

A 0 grams
B 25 grams
C 50 grams
D 60 grams

◆ Examine the Question
◆ Recall What You Know
◆ Apply What You Know

OBJ. 3 5.7 (C)

14 A student fills an ice cube tray with water and puts it into the freezer. What happens when the water in the ice cube tray freezes?

F Liquid → gas   H Solid → liquid
G Gas → solid   J Liquid → solid

OBJ. 3 5.7 (A)

15 What does a student learn when she measures the mass of her soccer ball?

A How fast it is moving
B How much heat it is giving off
C How much matter it contains
D How much space it takes up

OBJ. 3 5.7 (A)

16   **Jasmine made four cups of water. She added a different material in each cup and stirred it. Which of the following will dissolve in the cup?**

F   Salad oil   H   Sand   OBJ. 3
G   Sugar   J   Iron filings   5.7 (C)

17   **Which of these is the best conductor of heat?**

A   A rubber glove   C   A copper pot   OBJ. 3
B   A plastic spoon   D   A wooden handle   5.7 (A)

18   **What process is illustrated in the diagram?**

F   Boiling   OBJ. 3
G   Melting   5.7 (D)
H   Freezing
J   Burning

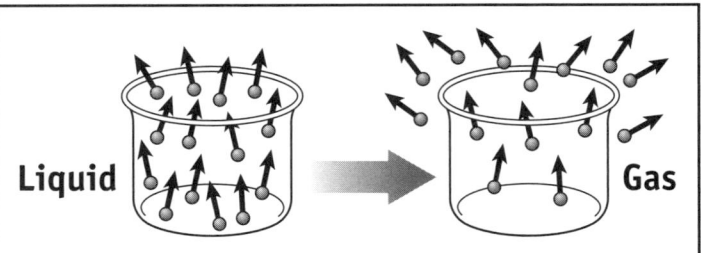

In an experiment, two students decide to test whether volume affects boiling point. Each students heats up different amounts of water in a container over a hot plate. Then they measure the temperature of the water when it begins to boil.

	Volume	Boiling Point
Student A	1 liter	100° C
Student B	2 liters	?

19   **What is the boiling point for the water heated by Student B?**

A   0° C   C   100° C   OBJ. 3
B   50° C   D   200° C   5.7 (D)

20   **In an experiment, sugar and iron filings are mixed together. What would be the fastest way to separate the iron filings from the sugar?**   OBJ. 3 5.7 (D)

F
- stir the mixture in water
- pour the wet mixture through a filter

H
- put the mixture in a beaker
- heat the beaker over a hot plate

G
- pour the mixture on a piece of paper
- use a pin to separate the filings

J
- stir the mixture in water
- boil the mixture

CHAPTER 8

MOTION, FORCE AND ENERGY

Energy has the ability to move matter. In this chapter, you will learn about motion, force, and energy and how they are related.

— MAJOR IDEAS —

★ **Motion** consists of speed and direction.

★ When **force** is applied to an object, it changes its motion.

★ **Energy** is an ability to change or move matter. There are many different kinds of energy. These include heat, light, electricity, and chemical energy. Energy has the ability to change its form.

★ **Light** can be **reflected** or **refracted**.

★ **Electricity** flows in a circuit. Electricity can produce heat, light, sound and magnetic effects.

★ A **vibrating** object produces **sound**.

FORCE AND MOTION

In the last chapter, you studied matter. In this chapter, you will study what makes matter move and change.

Let's begin with **motion**, or how an object moves. Motion occurs when an object is changing its position. If you watched a race, you would see runners move from one place to another. The length they travel is known as **distance**.

SPEED

Speed is how fast an object moves. It is the distance the object travels in a specific amount of time. For example, a car may travel 10 kilometers an hour. This means that every hour, the car travels 10 kilometers. How many kilometers would that car travel in two hours? In 3 hours? Fill in the chart below:

Time	Distance Traveled
1 hour	★ 10 kilometers
2 hours	★ _____
3 hours	★ _____
4 hours	★ _____
5 hours	★ _____

You can turn this same information into a line graph. Complete the graph below:

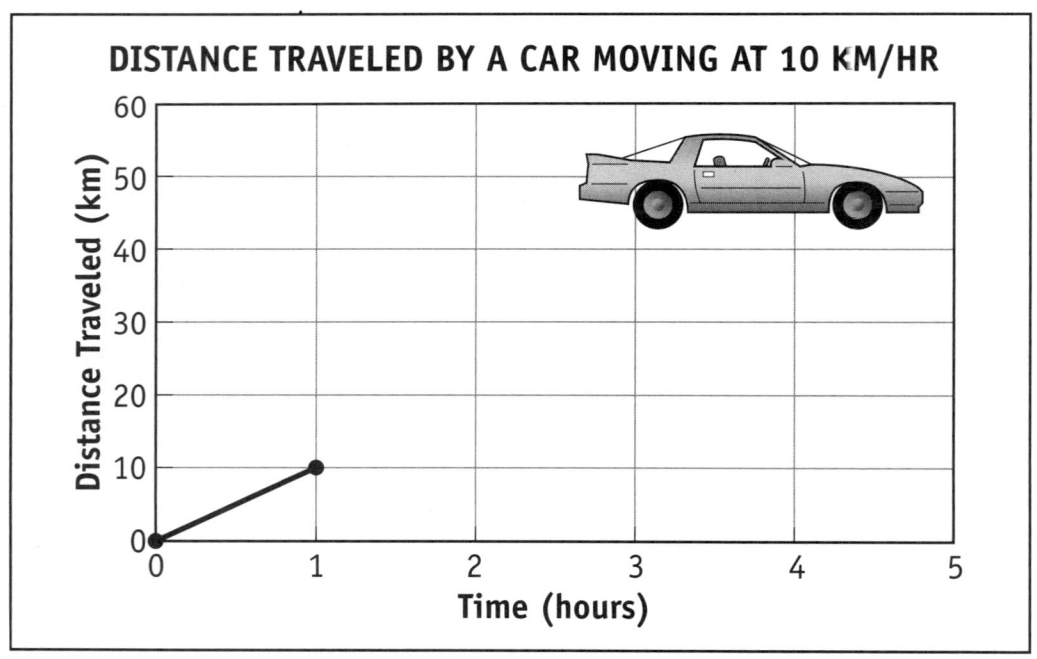

DISTANCE TRAVELED BY A CAR MOVING AT 10 KM/HR

DIRECTION

A second thing to consider when looking at motion is **direction**. If a car travels in the same direction for 2 hours at 50 km an hour, it will end up 100 km from its starting point. But if it travels for 1 hour and then turns back, at the end of two hours it will be back at the starting point. Can you explain why?

FORCE CHANGES MOTION

What makes objects move? Have you ever kicked a soccer ball resting on the ground? If so, you have seen the effect that force has on an object. When you kick a soccer ball, you apply a force that makes the ball start to move. The ball begins rolling, but at some point it will slow and stop. The ball stops rolling because of **friction**, the rubbing of the ball against the ground. If there were no friction to slow down the ball, the ball would keep rolling forever — until some other force stopped it.

A foot applies force to a soccer ball.

All changes in movement are caused by force. To start an object moving, whether a soccer ball or car, requires some force. **Force** is what pushes or pulls the object, causing it to change its speed or direction. **Gravity** is a force that pulls things towards the Earth. Once moving, things keep moving until some other force changes or stops their motion.

In outer space there is no gravity or friction. Moving objects just keep moving at the same speed and direction until they come into contact with some other force. Some outside force is needed to cause the object to slow down, speed up, or change its direction.

APPLYING WHAT YOU HAVE LEARNED

Describe what would happen to a rocket ship traveling in outer space that is moving at 100 km per second if no force is applied to it: _____

_____

_____

_____

FRICTION

Friction is the force that occurs when two things rub against each other. As two objects slide against each other, they grind and drag against each other. This is where friction comes from. For example, if you rub your hands together you create friction.

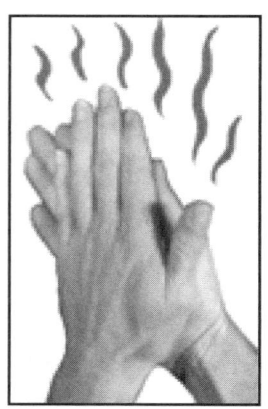

On Earth, moving objects encounter friction from the air and ground. The force of this friction slows moving objects down and eventually leads them to stop, unless force is applied to keep them moving.

APPLYING WHAT YOU HAVE LEARNED

When a car moves ahead, the force from the wheels moving the car forward is greater than the opposing force of friction — the tires rubbing against the roadway.

Identify two other examples from everyday life in which force leads to changes in an object's speed or direction.

A. _____

B. _____

WHAT FRICTION DEPENDS ON

The amount of friction depends on the type of surfaces that are rubbing against each other. When surfaces are smooth, they move against each other more easily. **Friction** is greater between rough surfaces. In order to reduce friction in the moving parts of machines, scientists and engineers use oil and other lubricants. These lubricants act to make the surface smoother, therefore causing less friction.

IS FRICTION GOOD OR BAD?

Friction is not all bad. In fact, friction can be very helpful. If you have ever tried to run on a wet floor, you know that too little friction can make you slip and slide. You need friction to get a grip. Without friction, we would be unable to walk, sit in a chair, or climb up stairs. Cars would be unable to be driven and machines would not work. Too much friction, however, will wear down moving parts.

APPLYING WHAT YOU HAVE LEARNED

✦ Which has more friction, a roadway before or during a heavy rainfall? Explain. _____

_____

FORCE, MASS AND GRAVITY

How much the speed of an object will change when force is applied depends on the amount of force and the size of the object. An object with **greater** mass requires **more** force to change its motion than an object with less mass.

APPLYING WHAT YOU HAVE LEARNED

✦ Think of a large boulder rolling down a hill. How much force do you think would be needed to stop it? _____

_____

_____

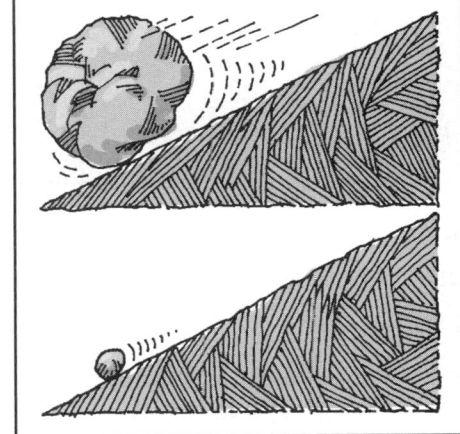

✦ Now think of a small pebble rolling down the hill at the same speed as the boulder. Do you think more or less force would be needed to stop the pebble than the boulder? _____

Explain. _____

_____

_____

Gravity. Why do objects fall to the ground? The Earth actually attracts objects with a force known as **gravity**. Because of this force, objects are pulled towards Earth's center. The force of gravity **increases** with the size of the object. Objects with greater mass are pulled to the Earth with greater force.

APPLYING WHAT YOU HAVE LEARNED

◆   Which object is pulled to Earth by a greater gravitational force — a paper clip or a wooden table? Explain your answer. _____

_____

_____

◆   Do similar objects with different masses fall at the same speed or at different speeds? Actually, they all fall at the same speed. Can you explain why? _____

_____

_____

ENERGY

Where does force come from? Force is always created by some kind of energy. Energy is harder to think about than matter. You cannot always see energy, but you can sometimes feel it. Some days you may feel more energetic than others. You feel like doing things. **Energy** is an ability to cause changes in matter. There are many kinds of energy. These include:

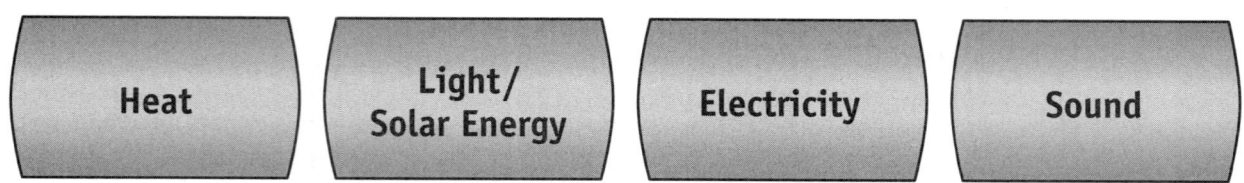

| Heat | Light/ Solar Energy | Electricity | Sound |

Let's take a closer look at these forms of energy.

HEAT

Heat is caused by the movement of tiny particles in matter. As the particles in an object move around more quickly, the object heats up. **Temperature** measures how fast these particles are moving. Water particles are moving faster at 100°C than at 50°C.

Heat can pass from one object to other objects. The moving particles in one piece of matter bump into the particles of neighboring materials and speed them up. Heat transfers from hotter objects to colder ones. For example, if you stirred a pot of soup on the stove with a metal spoon, after a few minutes the metal spoon would heat up. The heat energy passes from the hot soup to the spoon.

LIGHT

Light is another form of energy. Light comes from the energy given off by matter when tiny particles that make up matter collide with each other or are very hot. Light energy can move through water, air, and glass. Light can even move through empty space. **Solar energy** is a form of light energy. This light energy passes through millions of miles of space to reach the Earth. The next time you see a pretty sunrise or sunset, think about the fact that the light you see has left the sun eight minutes earlier. This is because it takes that long for this light to reach the Earth.

When light travels through air, water or space it travels in a straight line — remaining the same along its entire path. Light can pass through some materials, but not through others. Materials such as black construction paper just absorb light. Other materials reflect light. The light rays just bounce off these materials. Still others refract light. The light passes through these materials, but its rays will bend. Some materials absorb some light rays and reflect others. A red scarf absorbs some light rays but reflects red light.

APPLYING WHAT YOU HAVE LEARNED

How will the following materials affect light?

Materials	Lets it Pass Through	Absorbs Light	Reflects Light	Refracts Light
Water in a calm lake				
Pair of eyeglasses				
Shiny silver car				
Pair of blue jeans				

REFLECTION

Reflection involves the bouncing of a light ray off a surface. An incoming ray of light hits the object and bounces away. A mirror uses a shiny, smooth surface to reflect light. If a ray of light hits a mirror, it will bounce right back off.

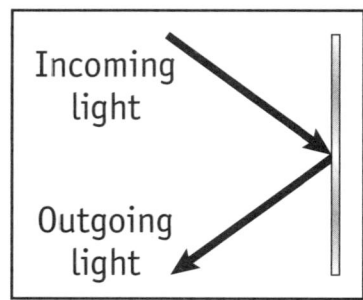

REFRACTION

Have you ever stuck a straight stick into a pool of water? If you have, you probably noticed that the stick appears to bend where it enters the water. This sometimes happens when light passes from one substance through another. The light does not travel at the same speed in the second substance. This makes the rays of light appear to bend. The bending of the light rays passing from one substance to another is called **refraction**. It often occurs when light passes into a transparent material. For example, you can see this when you dangle your feet in a pool or put a pencil in a glass of water.

APPLYING WHAT YOU HAVE LEARNED

Explain what happens when light shines into a glass of water:

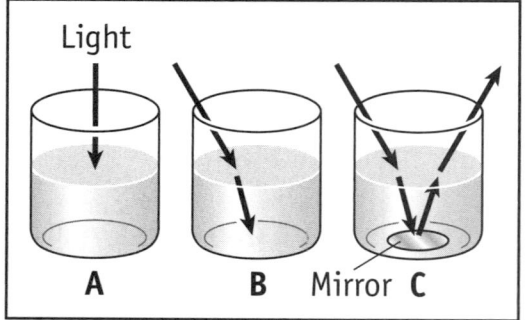

◆ **Figure A.** Light shines down into a glass container of water. The light strikes the water straight on. Explain what will happen to the light as it reaches the bottom of the container.

_____

◆ **Figure B.** Light shines into the container of water at an angle. As the light enters, explain how the beam of light appears at the surface of the water.

_____

◆ **Figure C.** A mirror is placed at the bottom of a container. Light shines into the container at an angle. Explain what happens to the light when it hits the mirror at the bottom of the container.

_____

LENSES

Lenses take advantage of the ability of some materials to bend, or **refract**, light.

A **lens** is a piece of glass or other material that is curved. A **convex** lens bends light rays towards each other. A **concave** lens bends them apart. Eyeglasses use lenses to correct people's vision.

A telescope uses lenses to focus light rays from faraway places so that they appear more clearly. The front of your eye acts as a human lens. It takes light and focuses it on the back of the eye, forming an image that is sent to your brain.

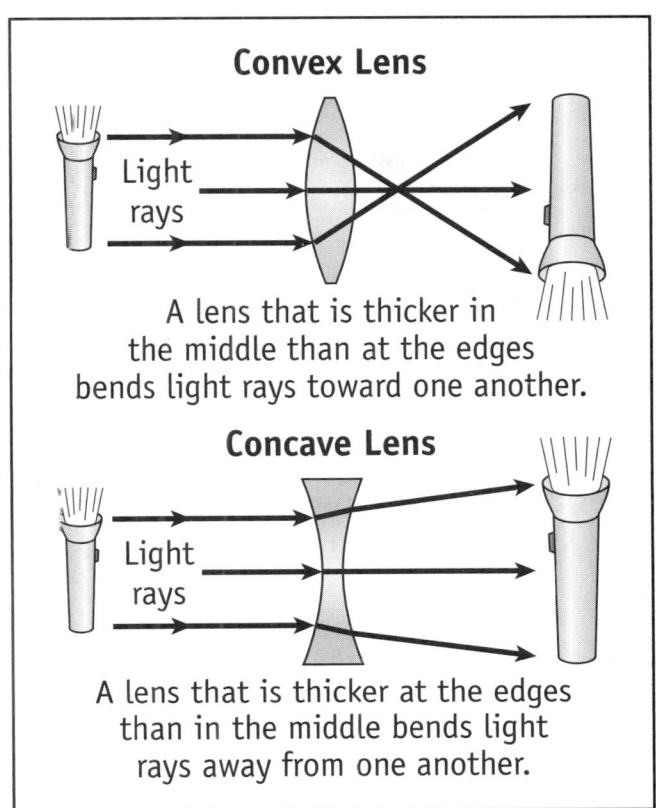

Convex Lens

A lens that is thicker in the middle than at the edges bends light rays toward one another.

Concave Lens

A lens that is thicker at the edges than in the middle bends light rays away from one another.

ELECTRICITY

Electricity is another form of energy. It is made by fast-moving charged particles. Each charged particle carries some energy. Electricity has many special properties.

Electricity can flow easily through many types of materials. Most metals are good conductors of electricity. So is water. However, not all materials conduct electricity. Wood and rubber do not conduct electricity

APPLYING WHAT YOU HAVE LEARNED

Describe two ways that you used electricity today.

1. _____

_____

2. _____

_____

ELECTRIC CIRCUITS

Electricity can flow in a circuit. A typical **circuit** is a system with several parts:

| A source that provides electricity | Something that uses electricity | Wires to carry the electricity back and forth |

A battery is one type of source that produces electrical energy. Every battery has a positive (+) and a negative (−) side. Wires connected to the battery can carry the electricity.

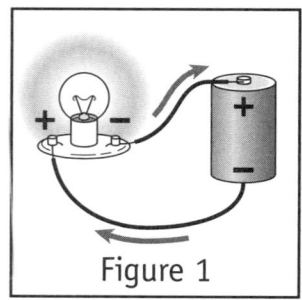

Figure 1

However, the electricity will only move if the wires and other parts form a complete circuit. This gives a continuous path for the particles to move through.

For example, imagine a simple circuit with a light bulb at one end and a battery at the other (Figure 1) . Electric current leaves one side of the battery and moves through wires to the light bulb. The electricity moves through the light bulb. The light bulb uses the electrical power to light up. The electric current continues moving through the wires back to the battery.

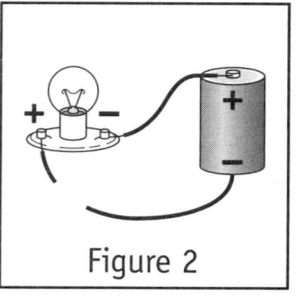

Figure 2

If the circuit is cut at any point, the electricity moving around the circuit will stop (Figure 2). As a result, the light will have no electricity flowing into it, and the bulb will go out.

The wires must also be connected to each side of the battery. If both ends of the wire are connected to the same side of the battery, the electricity will not leave the battery to make a circuit (Figure 3). That means that the electricity will not go around the wire, and the bulb will receive no power and go out.

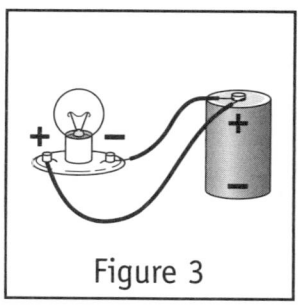

Figure 3

If both ends of the wire are connected to the same side of the light bulb, the electricity will go around the circuit without going into the light bulb (Figure 4). The bulb will not light up.

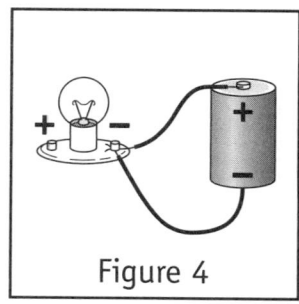

Figure 4

APPLYING WHAT YOU HAVE LEARNED

The circuits below have light bulbs, a battery, and wires.
In which illustration will the bulbs go on?
Explain why the other circuits will not light the bulbs.

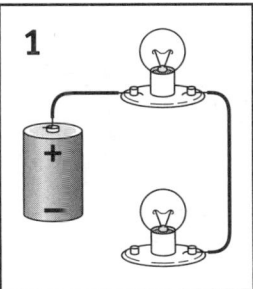 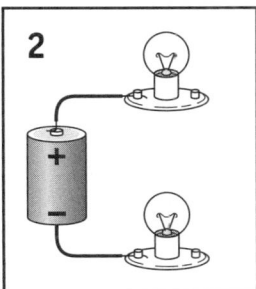 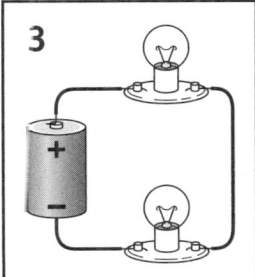 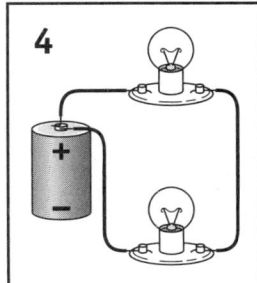

✦ What illustration shows a circuit capable of lighting the bulbs? _____

✦ Why won't the other circuits light the bulbs? _____

_____

_____

OTHER PROPERTIES OF ELECTRICITY

Electricity also has certain other properties. These properties involve magnetism, heat, light and sound.

★ **Magnetism.** When electricity runs through a wire, the wire becomes **magnetic**. A magnet made by electricity is called an **electromagnet**. When the electricity is turned off, the wire is no longer magnetic. This property of electricity makes it possible to make electric motors.

★ **Heat.** When electricity runs through some materials, it makes them hot. Irons, waffle makers, toasters, and electric heaters all produce heat when electricity runs through them.

★ **Light.** When electricity runs through some materials, it makes them so hot they glow. Because of this property, electricity makes light bulbs work. Electricity also causes flashes of lightning in the sky.

★ **Sound.** Electricity is also capable of making sounds. Televisions, radios, and stereo speakers use electricity to make sounds. Electricity in nature causes the loud sounds of thunder.

APPLYING WHAT YOU HAVE LEARNED

Examine the electrical appliances illustrated below. Identify what kind of energy each item produces.

Lightbulb	Toaster	Door bell	Radio

SOUND ENERGY

Sound is created by vibrating objects. Some force must be applied to make the object vibrate. For example, when a person pulls a guitar string, the string begins to vibrate. The energy from the vibrating string causes the air to vibrate. Our ears are very sensitive to these vibrations. The energy from these vibrations passes to our ears, and we hear these vibrations as sound.

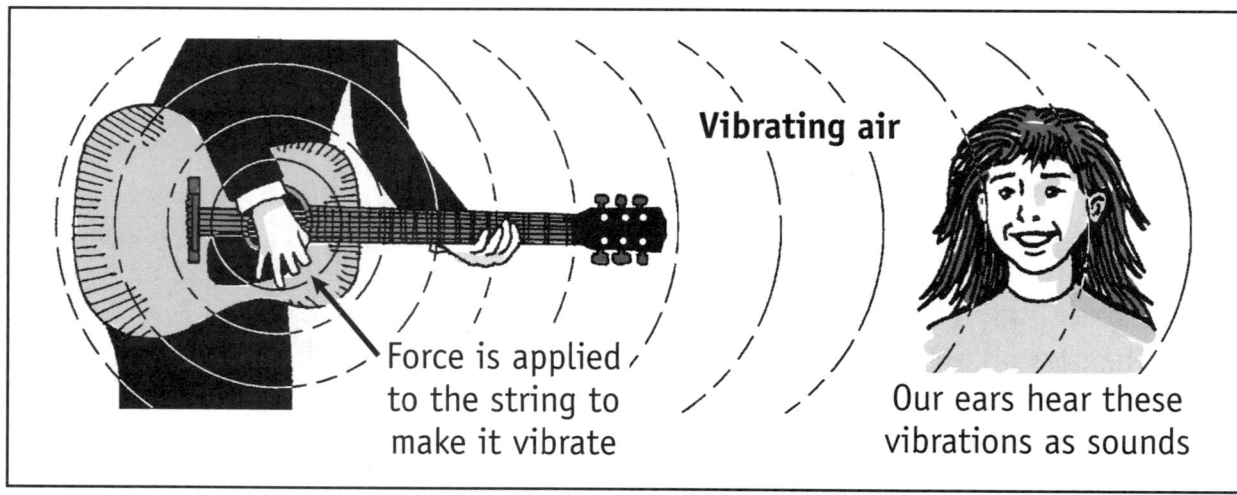

Vibrating air

Force is applied to the string to make it vibrate

Our ears hear these vibrations as sounds

Most sounds we hear travel through the air. But sound can also travel through many forms of matter, including liquids and solids. Sound actually travels faster through many solids than through the air. The vibrations from a vibrating object spread out in waves. These sound waves carry energy from the sound source outward in all directions.

APPLYING WHAT YOU HAVE LEARNED

1. How can we test whether vibrating objects produce sound?_____

    _____

2. Sometimes, if you put your ear on the wall it is possible to pick up what is
    being said in the next room. Why is this so?_____

    _____

3. Can sound travel through empty space? _____ Why or why not?

    _____

    _____

ENERGY CAN CHANGE ITS FORM

At the beginning of this chapter, you learned that energy can change its form. Now that you know more about the different kinds of energy, you can see how this works:

★ **Electrical Energy → Light Energy.** Electricity runs through a material that gets very hot and glows. For example, it can get so hot that it can make a light bulb glow. In this example, *electrical energy* became *light energy.*

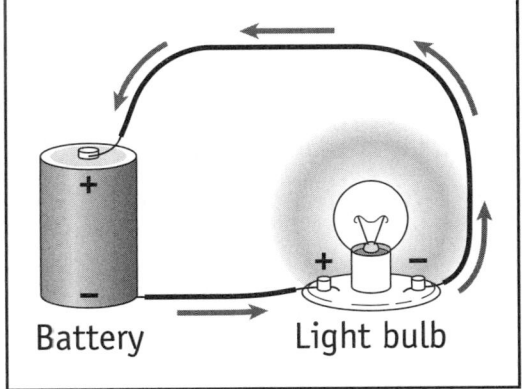

Battery   Light bulb

★ **Solar Energy → Heat Energy.** We use the sun's energy in different ways almost every day. For example, anyone who has ever hung laundry outside to dry is using the sun's heat to dry clothes. If you go to the beach and lie in the sun, after a few minutes you will start to feel hot. The energy that comes from the sun, *solar energy*, is heating your body.

★ **Solar Energy → Chemical Energy.** Through photosynthesis, plants change energy from the sun into *chemical energy* stored in food.

WHAT YOU SHOULD KNOW

A. You should know that **motion** consists of speed and direction. When **force** is applied to an object, it changes its motion.

B. You should know that **energy** has the ability to change or move matter. There are many different kinds of energy. These kinds of energy include heat, light, electricity, and chemical energy. Energy can change its form.

C. You should know that **light** can be **reflected** or **refracted**.

D. You should know that **electricity** flows in a circuit. Electricity can produce heat, light, sound and magnetic effects.

D. You should know that a **vibrating** object produces **sound**.

CHAPTER STUDY CARDS

Motion

Motion occurs when an object is changing its location. Motion includes speed and direction.

★ **Speed**. Speed measures the distance an object travels in a given amount of time, often measured in km/hr.

★ **Direction** is the path or route an object takes.

Force

Force is what pushes or pulls an object to make it change its speed or direction. An object with greater mass requires more force to change its motion than one with less mass.

★ **Friction** is a force created by the rubbing of two surfaces. On Earth, friction slows down moving objects.

★ **Gravity**. The force pulling objects to the Earth. The force of gravity increases with the size of the object.

Energy

Energy has the ability to change matter. There are many forms of energy:

★ **Electricity**. Energy of charged particles.

★ **Heat**. When the energy of particles in matter move, they heat up.

★ **Light**. Can travel through space and many materials.

★ **Solar Energy**. Light from the sun.

★ **Sound Energy**. Energy carried by sound waves.

Light

★ When light travels through space, air, or water, it travels in a straight line.

★ Some materials absorb light.

★ **Reflection**. Some materials reflect light. The light rays bounce off these materials

★ **Refraction**. Light rays become bent as they pass through one material to another.

CHECKING YOUR UNDERSTANDING

1 Jack arranged some wire, a battery and a small light bulb in four different ways as shown below. Which one will light up when the switch is closed?

OBJ. 3
5.8 (C)

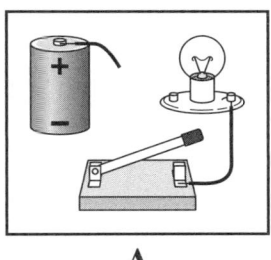
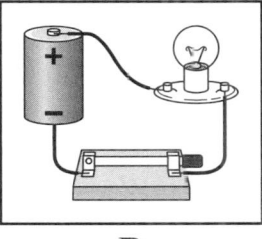
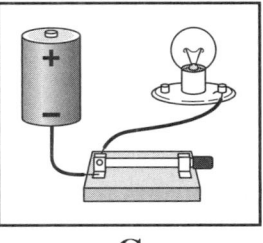
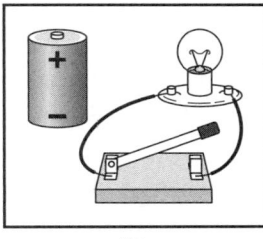

A   B   C   D

HINT To answer this question correctly, you must understand electric circuits. In a circuit, electricity leaves the source and travels around a pathway until it returns to the other side, or pole, of the source. Only **Choice B** will create a closed circuit when the switch is closed.

*Now try answering some
additional questions on your own:*

2 If the person in the picture lets go of the rope, the weight (w) will fall to the ground. What force is acting to pull the weight to the ground?

   F Magnetism
   H Friction
   G Gravity
   J Electricity

OBJ. 3
3.6 (A)

3 A student measures the time it took her to run 50 meters. What is she able to calculate using her distance and time measurements?

   A Mass   **C** Speed
   B Force   **D** Direction

OBJ. 3
3.6 (A)

4 Which tools are needed to measure the speed of a sliding penny?

   F Stopwatch, ruler   **H** Thermometer, stopwatch
   G Balance, ruler   **J** Thermometer, balance

OBJ. 3
3.6 (A)

5   A student watches cars race around a track. What information can she best use to describe the motion of the cars?

   A   Shape of the cars
   B   Temperature of the tires
   C   Speed of the cars
   D   Mass of the cars

OBJ. 3
3.6 (A)

6   Sound is created by a finger pulling on a guitar string. This is because the force on the string causes —

   F   heat   H   vibrations
   G   magnetism   J   electricity

OBJ. 3
5.8 (D)

7   What force slows down a rolling baseball?

   A   Magnetism   C   Light
   B   Heat   D   Friction

OBJ. 3
3.6 (A)

8   A student is out for a walk and sees a flash of lighting. What does this flash of lightning show?

   F   Light can be reflected.
   G   Electricity flows in a circuit.
   H   Light is a form of heat.
   J   Electricity can produce light.

OBJ. 3
5.8 (C)

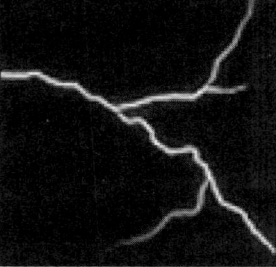

9   Light that bounces off a smooth, shiny surface is —

   A   refracted
   B   reflected
   C   absorbed
   D   transmitted

♦ Examine the Question
♦ Recall What You Know
♦ Apply What You Know

OBJ. 3
5.8 (B)

10   Which of the following converts electrical energy into heat?

   F   An electric fan   H   An electric stove
   G   A light switch   J   A CD player

OBJ. 3
5.8 (A)

11   Electricity traveling through a wire is an example of —

   A   force being applied by a simple machine
   B   energy flowing through the water cycle
   C   energy being transferred from one place to another
   D   the sun's rays causing heat

OBJ. 3
5.8 (C)

12 **What two forms of energy does the fire from these burning candles release?**

F   Light and heat energy
G   Sound and chemical energy
H   Magnetic and light energy
J   Electrical and magnetic energy

OBJ. 3
5.8 (A)

13 **A student lives near an airport. Whenever a plane flies over his house, his windows shake. Which of these scientific facts best explains why this happens?**

A   Sound waves transmit energy.
B   Heat waves transmit energy.
C   Radio waves transmit energy.
D   Electromagnetic waves transmit energy.

OBJ. 3
5.8 (D)

14 **This picture shows the path of a fruit fly. How much distance did the fruitfly travel to go from Point A to Point D? Use a ruler to measure the distance to the nearest centimeter. Record and bubble your answer in the grid.**

OBJ. 3
3.6 (A)

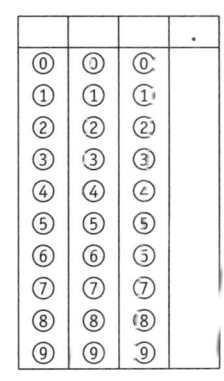

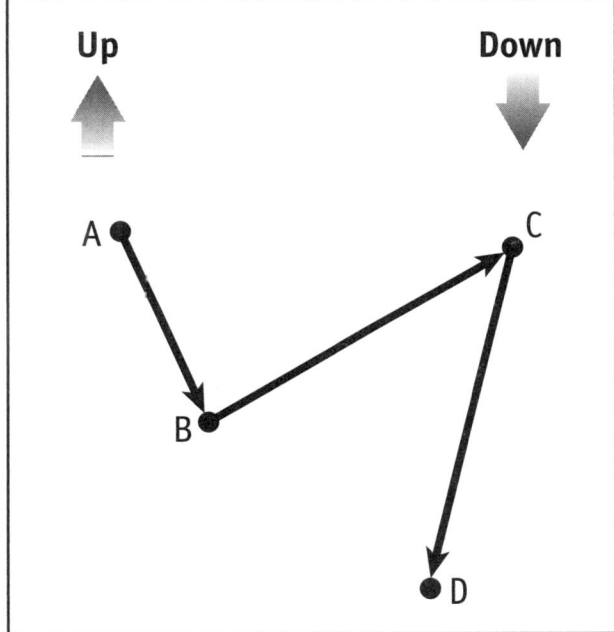

15 **The instrument on the right is used to see which materials can conduct electricity. Which group contains items that could all conduct electricity to complete the circuit?**

A   rubber ball, plastic comb, nail
B   paper clip, penny, iron screw
C   cork, dollar bill, tweezers
D   pencil, eraser, spoon

OBJ. 3
5.7 (A)

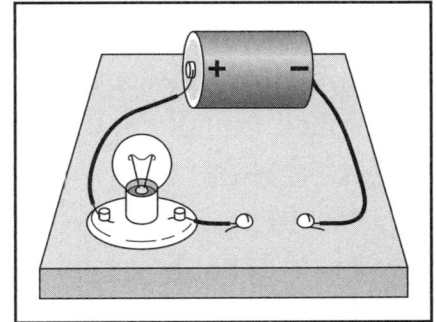

16 **Which of the following would reflect rather than refract light?**

    F A magnifying glass

    G A lens

    H A mirror

    J A glass of water

OBJ. 3
5.8 (B)

17 **Blowing on a trumpet will make a sound because the —**

    A trumpet heats the air

    B air in the trumpet vibrates

    C trumpet cools the air

    D trumpet causes air to refract

OBJ. 3
5.8 (D)

Libby went on a trip with her family. In her log, she recorded the time of day and how long her family had traveled since they started.

Time of Day	Total Distance Traveled
8:00 AM	0 km
9:00 AM	30 km
10:00 AM	60 km
11:00 AM	90 km

18 **What speed is Libby's family traveling in kilometers/ hour? Record and bubble your answer in the grid.**

OBJ. 3
3.6 (A)

19 **The graph to the right shows the movement of a car during a 4 hour trip. What was the speed of the car, if it traveled at the same speed for all four hours?**

    A 0 km/h

    B 50 km/h

    C 100 km/h

    D 200 km/h

OBJ. 3
3.6 (A)

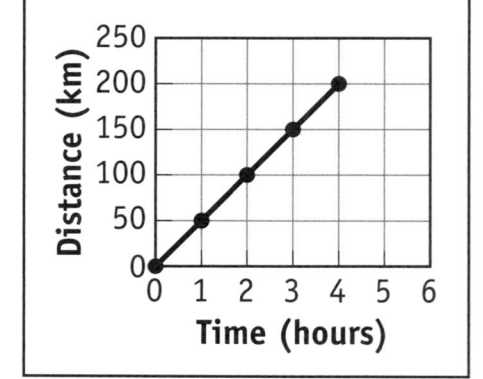

Marble	Tennis ball	Wooden box	Small boulder
2 g	10 g	1 kg	100 kg

20 Which object would require the most force to move a distance of 5 meters?

F Marble

G Tennis ball

H Wooden box

J Small boulder

OBJ. 3
3.6 (A)

21 Which of these appliances changes electrical energy into light energy?

OBJ. 3
5.8 (C)

   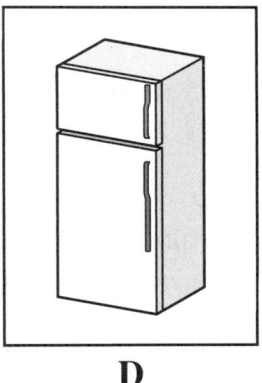

A   B   C   D

22 Which of the following shows light being reflected?

OBJ. 3
5.8 (B)

   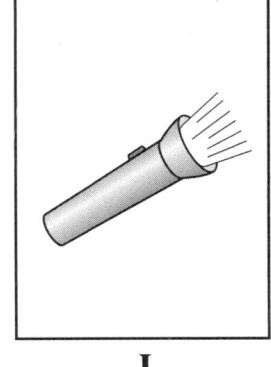

F   G   H   J

CHECKLIST OF OBJECTIVES IN THIS UNIT

Directions. Now that you have completed this unit, place a check (✔) next to those objectives you understand. If you are having trouble recalling information about any of these objectives, review the chapter listed in the accompanying brackets.

☐ You should be able to differentiate among forms of energy including light, heat, electrical, and solar energy. **[Chapter 8]**

☐ You should be able to identify and demonstrate everyday examples of how light is reflected, such as from tinted windows, and how it is refracted, such as in cameras, telescopes, and eyeglasses. **[Chapter 8]**

☐ You should be able to demonstrate that electricity can flow in a circuit and can produce heat, light, sound, and magnetic effects. **[Chapter 8]**

☐ You should be able to verify that vibrating an object can produce sound. **[Chapter 8]**

☐ You should be able to classify matter based on its physical properties including magnetism, physical state, and the ability to conduct or insulate heat, electricity, and sound. **[Chapter 8]**

☐ You should be able to demonstrate that some mixtures maintain the physical properties of their ingredients. **[Chapter 7]**

☐ You should be able to identify changes that can occur in the physical properties of the ingredients of solutions, such as dissolving sugar in water. **[Chapter 7]**

☐ You should be able to observe and measure characteristic properties of substances that remain constant, such as boiling points and melting points. **[Chapter 7]**

☐ You should be able to measure and record changes in the position and direction of the motion of an object to which a force such as a push or pull has been applied. **[Chapter 8]**

UNIT 4

SPACE AND EARTH SCIENCES

In this unit, you will review what you need to know about the Space and Earth Sciences. You will learn about the sun and its solar system, including the nine planets and the moon.

Then you will learn about our planet the Earth. You will read about the forces that build up and tear down the Earth's surface, and about Earth's resources.

An artist's drawing of the planets as they revolve around the sun.

★ **Chapter 9: Space and the Solar System**
In this chapter, you'll learn the characteristics of a star, including the sun. You will learn about our solar system. Finally, you will learn how Earth's movements around the sun cause day and night and the change of seasons. You will also learn about the moon's movements.

★ **Chapter 10: Our Changing Earth**
In this chapter, you will learn about the forces that build up and tear down Earth's surface, such as earthquakes, erosion, and weathering. You will learn how our planet consists of different materials, including rock, water, soil and gas. Finally, you will learn how scientists use tree rings and rocks to discover the Earth's past history.

★ **Chapter 11: Earth's Resources, Interactions and Cycles**
In this final chapter, you will learn about Earth's renewable, non-renewable, and inexhaustible resources. You will learn about the properties of soil, the effects of the oceans and weather, and the water, nitrogen and carbon cycles.

CHAPTER 9

SPACE AND THE SOLAR SYSTEM

In this chapter, you will learn about our solar system, including its nine planets and their positions in relation to the sun.

— MAJOR IDEAS —

★ **Stars** are enormous balls of superheated gases.

★ The **sun** is a star. It is the major source of energy for the Earth.

★ The **solar system** consists of the sun and nine planets — Mercury, Venus, Earth, Mars, Jupiter, Saturn, Uranus, and Neptune.

★ The Earth spins or **rotates** on its axis. This spinning causes us to have day and night. The Earth is **tilted** on its **axis** as it revolves around the sun. This explains why the seasons of the year change from spring and summer to fall and winter.

★ The moon orbits the Earth. The sun's light reflects off the moon, causing us to see the moon at night.

OUR SOLAR SYSTEM

Most of the universe is empty space. In this vast empty space are **galaxies** — groups of stars. Our sun is one of these stars. The sun is the center of our solar system. Our planet, Earth, is also part of this vast system.

STARS

Stars are enormous balls of superheated gases. The nearest star to Earth, after the sun, is 40 trillion kilometers (*25 trillion miles*) away. For this reason, scientists use very powerful telescopes to see and study stars.

Stars are formed by clouds of gases and dust in space. These gases and dust come together and begin to spin. Eventually, they form a star. Each star actually acts as a giant nuclear reactor.

The center of a star is extremely hot and dense. The star's tremendous pressure causes tiny particles of matter to join together. This process releases tremendous amounts of energy which move outward from the star's center. This energy finally radiates from the

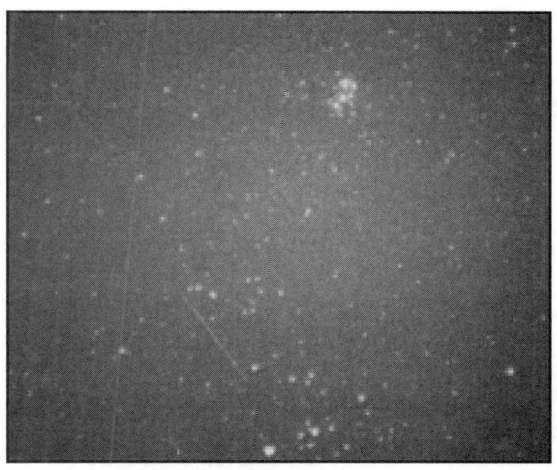
The night sky with stars sparkling

star's surface across space at the speed of light. Stars can last for billions of years. Although stars move in space, they appear steady when we look at the sky because of their great distances from Earth.

THE SUN

The **sun** is a star. It is one of millions of stars in the galaxy we know as the **Milky Way**. It is the closest star to Earth. Like other stars, the sun was formed by hot gases and dust in space that pulled together and started to spin. The sun's reactions produce enormous amounts of energy, giving off light and heat that travels through space.

The sun is the source of most of our energy on Earth and the rest of the solar

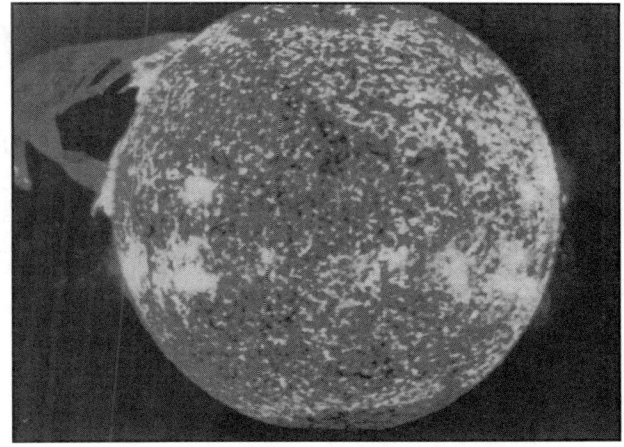
The sun, a hot ball of gases

system. It is the source of our heat and allows living things on our planet to survive. The sun's energy is the main influence on Earth's climate and weather.

The sun is the largest object in our solar system. It is an extremely hot ball of gases — about 15 million degrees Celsius. Each square centimeter of its surface provides as much light as a 6,000 watt lamp. Cooler areas of the sun's surface appear to us as dark **sunspots**.

THE PLANETS

Planets are objects of rock, metal, ice and gas that circle the sun. They do not give off their own light as the

sun does. There are nine planets in our solar system. They range from tiny rocky planets to huge gas giants with rings. The planets circle the sun. The eight planets, in their order from the sun, are: Mercury, Venus, Earth, Mars, Jupiter, Saturn, Uranus, and Neptune. An easy way to remember the order of the planets is to memorize this sentence: **My Very Excellent Mother Just Sent Us Nachos.** The first letter of each word represents a planet in its correct order.

The largest planet is Jupiter — ten times the width of Earth. It is so large that all of the other eight planets could fit inside it. The tiniest planet is Mercury. In general, the farther away a planet is from the sun, the cooler it is.

APPLYING WHAT YOU HAVE LEARNED

This table shows that Earth is 149.6 million kilometers from the sun (*93 million miles away*). Suppose there was an imaginary road from the Earth to the sun. A car traveling 100 km an hour would take 1,496,000 hours (*about 170 years*) to reach the sun. If it took 170 years to drive from Earth to the sun, about how many years would it take to drive from Jupiter to the sun at the same speed?

DISTANCE OF PLANETS FROM THE SUN

Planet	Distance from the Sun
Mercury	57.9 million km
Venus	108.2 million km
Earth	149.6 million km
Mars	227.9 million km
Jupiter	778.3 million km
Saturn	1,427.0 million km
Uranus	2,871.0 million km
Neptune	4,497.1 million km

Answer: _____

APPLYING WHAT YOU HAVE LEARNED

Look at the diagram of the solar system below. The planets have not been labeled. Name each of the planets in the order of their distance from the sun.

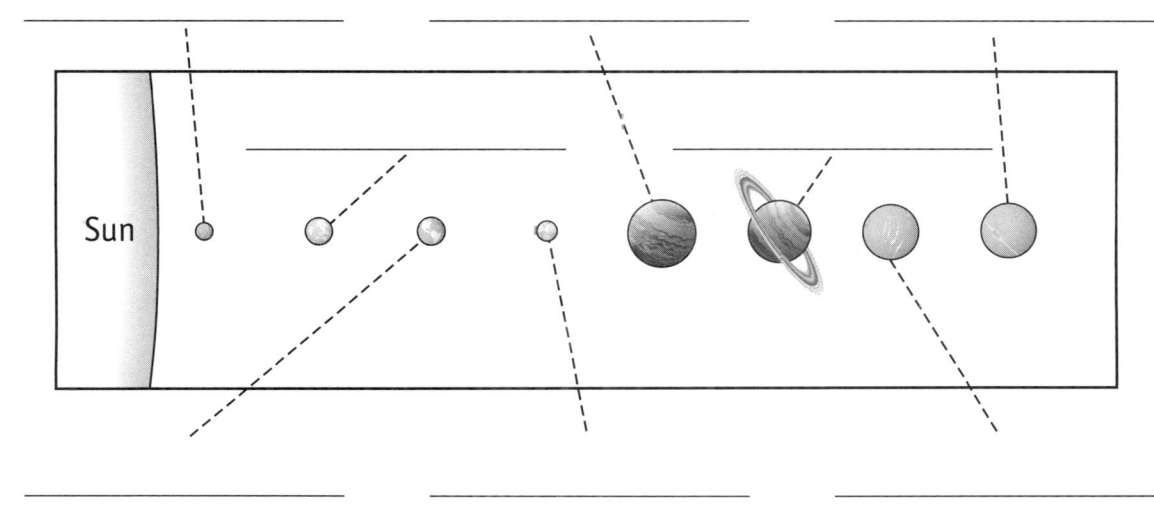

The diagram above gives you some idea of the positions of the planets in relationship to the sun. However, the diagram is **not** drawn to scale. If you look at the table on page 130, you would see that Neptune is more than 90 times farther from the sun than Mercury. A scale model of the solar system would therefore have to show Neptune 90 times farther away than Mercury. Unfortunately, such a model would never fit on this page.

In order to see the entire solar system on a single page, the distance between the planets is often changed. The same is true of their sizes. The sun is much larger than any planet. Its width across is 10 times that of Jupiter, the largest planet. The sun is 100 times wider than Earth. As a result, the size of the sun and planets on this map are also not shown to scale. Again, this is done so that the entire solar system can be viewed on a single page.

APPLYING WHAT YOU HAVE LEARNED

✦ What changes might you introduce to make this model of the solar system more accurate? _____

_____

_____

You can see some of the planets through a telescope. Mars has a red surface. Jupiter has a giant red spot and several large moons. Saturn is famous for its rings. Jupiter, Saturn, Uranus and Neptune are the largest planets because they are mainly gas.

APPLYING WHAT YOU HAVE LEARNED

The farther a planet is from the sun, the longer it takes to complete its orbit. This is because it has a longer distance to travel. The force of **gravity** explains the movement of the Earth and other planets around the sun.

HOW LONG IT TAKES A PLANET TO ORBIT THE SUN

Planet	Time to Complete One Orbit
Mercury	87.9 Earth days
Venus	224.7 Earth days
Earth	365.0 Earth days
Mars	687.0 Earth days
Jupiter	11.9 Earth years
Saturn	29.5 Earth years
Uranus	84.0 Earth years
Neptune	164.8 Earth years

★ Which planet takes the longest to orbit the sun?_____

★ Why does it take longer?_____

_____

★ Which planet takes the least time to orbit the sun?_____

★ Which planet takes 30 times longer than Earth to orbit the sun?_____

Our solar system has other bodies besides the planets. **Comets** are made of ice, rocks and dust that circle the sun in long oval-shaped orbits. As a comet approaches the sun, some of its ice turns to gas, creating a giant, glowing tail.

THE MOVEMENT OF THE EARTH AND MOON

The planet **Earth** actually moves in two different ways at the same time: it **rotates** on its axis, and it **revolves** around the sun.

EARTH'S ROTATION

The Earth **rotates**, or spins, around its **axis** — an imaginary line running through the center of Earth from the North Pole to the South Pole. This rotation takes 24 hours, causing **day and night** on Earth. Night occurs on those parts of the Earth that are away from the sun's rays.

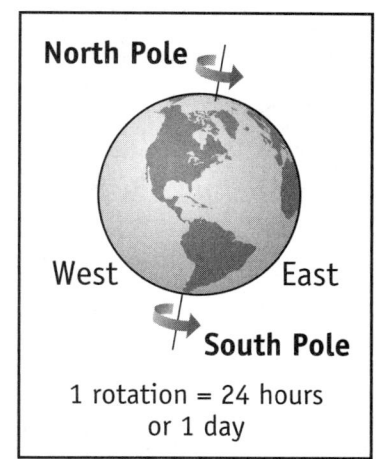

North Pole

West   East

South Pole

1 rotation = 24 hours or 1 day

EARTH'S REVOLUTION

The Earth also **revolves**, or circles, around the sun while it rotates on its axis. It takes just over 365 days (*one year*) for the Earth to complete one **revolution** around the Sun.

The Earth tilts on its **axis**. Because of this tilt, the sun's rays hit the Northern Hemisphere more directly in summer than in winter. When it is summer in the Northern Hemisphere, it is winter in the Southern Hemisphere. This is because the Southern Hemisphere tilts away from the sun and receives less direct sunlight in winter.

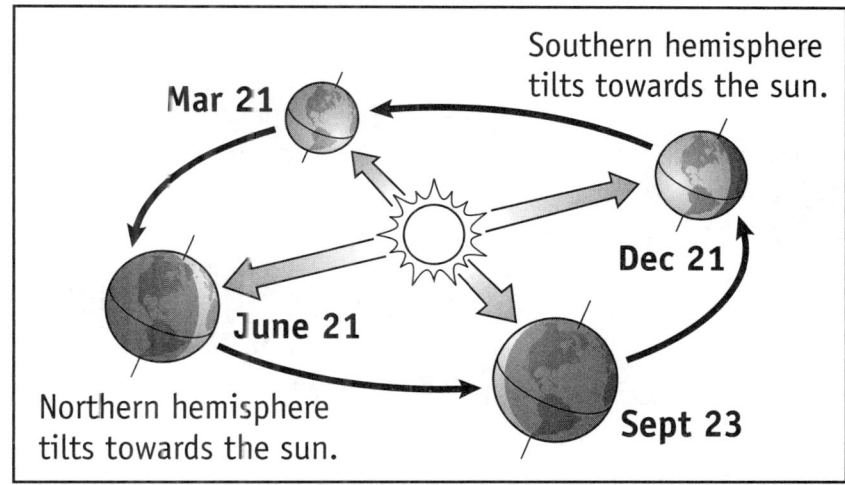

Mar 21

Southern hemisphere tilts towards the sun.

Dec 21

June 21

Northern hemisphere tilts towards the sun.

Sept 23

MOVEMENT OF THE MOON

The Earth has one satellite, the moon. It circles the Earth every $29\frac{1}{2}$ days. The moon is about one-quarter the size of Earth itself. Because of its smaller size, the moon's gravity is only a fraction of Earth's gravity. Objects of the same mass weigh much less on the moon than on Earth. The pull of the moon's gravity on the

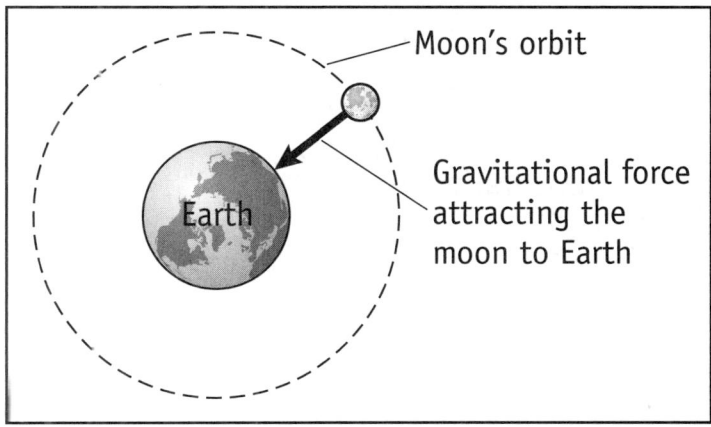

Moon's orbit

Gravitational force attracting the moon to Earth

Earth

Earth is the main cause of the rising and falling of ocean **tides**.

Have you ever walked home at night with the moon shining brightly? The moon does not produce its own light. The moon appears bright because it reflects light from the sun. You can think of the sun as a lightbulb, and the moon as a mirror that reflects light from that lightbulb.

Image of the moon taken by Apollo 11 astronauts

The amount of the moon's surface that shines each night changes over time in a cycle that repeats itself about once a month. Every $29\frac{1}{2}$ days, the moon appears as a thin sliver, grows into a crescent, expands into a full moon, and then becomes a crescent again. Eventually it narrows until it becomes completely dark in the night sky. These changes are caused by the moon's orbit and the different views from Earth of the sun's rays reflected on the moon.

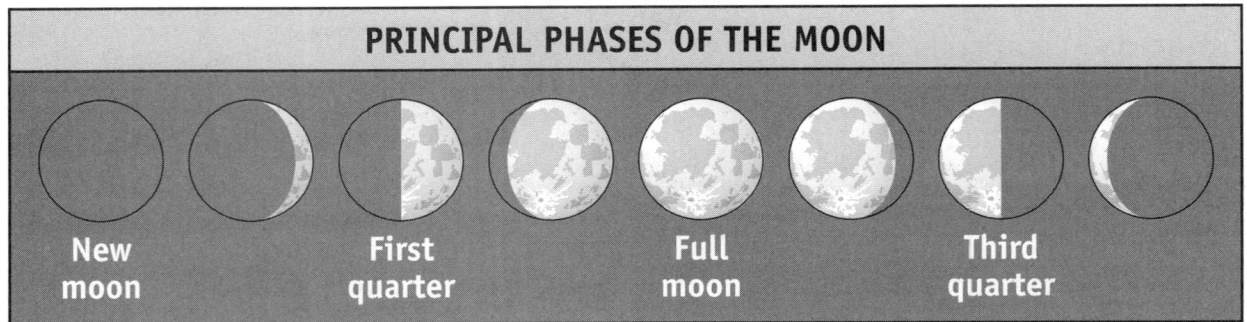

WHY DOES THE MOON APPEAR TO CHANGE?

When the Earth and sun are on the same side of the moon, the entire moon is lit up. However, when the moon is between the Earth and the sun, the moon appears to be totally dark.

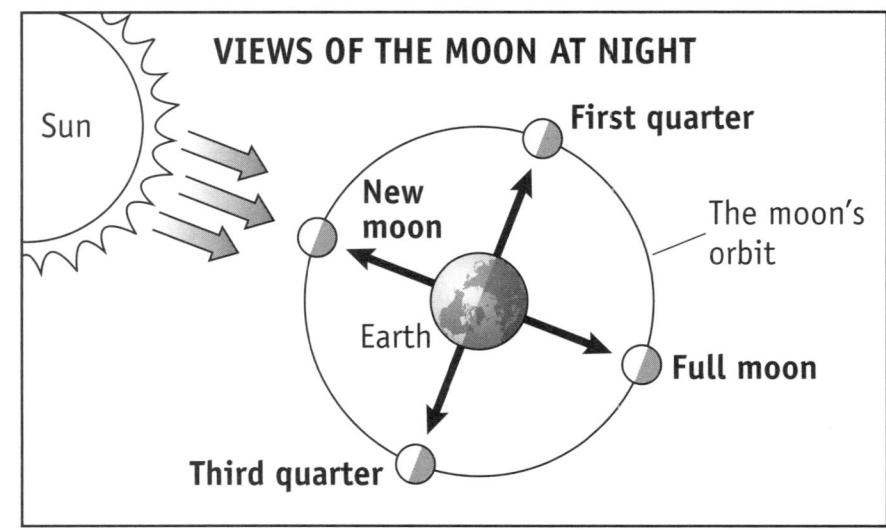

The moon is made of hard rock with craters. There is no water, air, soil or living things. The footprints left by the astronauts will last for centuries, since there's no wind on the moon. Since the moon has no atmosphere, there is no weather like we have on Earth. Meteors hit the moon without slowing down. When the moon faces the sun, it becomes extremely hot. Once the moon's surface is in shade, it becomes quite cold. Temperatures on the moon are extreme — ranging from 100°C to -173°C.

WHAT YOU SHOULD KNOW

A. You should know that **stars** are enormous balls of superheated gases. The **sun** is a star. It is the major source of energy for the Earth.

B. You should know that the **solar system** consists of the sun and eight planets — Mercury, Venus, Earth, Mars, Jupiter, Saturn, Uranus, and Neptune.

C. You should know that the Earth **rotates** (spins) on its axis. This spinning causes us to have day and night. The Earth is **tilted** on its **axis** as it revolves around the sun. This explains why the seasons of the year change from spring and summer to fall and winter.

D. You should know that the moon orbits the Earth every $29\frac{1}{2}$ days. It is the sun's reflection on the surface of the moon that causes us to see the moon at night.

CHAPTER STUDY CARDS

The Solar System

★ **Stars.** Stars are enormous balls of super-heated gases. They were formed by clouds of gases and dust in space.

★ **The sun.** The sun is a star. It is the largest body in the solar system. The sun is the source of most of our energy

★ **Planets.** There are 8 planets. Each planet orbits the sun. The eight planets in their order from the sun are: Mercury, Venus, Earth, Mars, Jupiter, Saturn, Uranus, and Neptune.

The Earth and Moon

★ **The moon** orbits the Earth every $29\frac{1}{2}$ days. It reflects the sun's light and appears on Earth in changing shapes or phases.

★ **Earth.** Earth's tilt on its axis gives us 4 seasons: spring, summer, fall and winter.

• The Earth rotates on its axis every 24 hours, creating night and day.

• The Earth revolves around the sun once every year.

CHECKING YOUR UNDERSTANDING

1 **Which of these would be the best model to show how the moon appears from Earth?**

OBJ. 1
5.3 (C)

A Lamp that is shut off   **C** Flashlight shining on a turning ball

B Street light that flickers   **D** Candle burning out

HINT To answer this question, you need to understand that a model represents something else. The moon does not give off its own light: it reflects light from the sun. Only **Choice C** provides an example of a round object reflecting light, so it is the best answer.

Now try answering some additional questions on your own:

PLANET DIAMETERS (IN KILOMETERS)

Mars	Venus	Earth	Jupiter	Saturn
6,794	12,104	12,756	142,800	120,660

2 **Which of these planets is most different in size from the Earth?**

F Venus   **H** Mars

G Jupiter   **J** Saturn

OBJ. 4
3.11 (C)

3 **Which best describes the movement of the moon and the Earth?**

A The moon orbits the Earth.

B The Earth orbits the moon.

C The sun revolves around the moon.

D The Earth and moon rotate with each other.

OBJ. 4
3.11 (C)

4 **The rotation of the Earth on its axis causes —**

F four seasons

G one year

H the moon's changing size

J day and night

◆ Examine the Question
◆ Recall What You Know
◆ Apply What You Know

OBJ. 4
5.6 (A)

5 Which of the following illustrations correctly depicts the movement of the Earth, sun and moon?

OBJ. 4
5.6 (A)

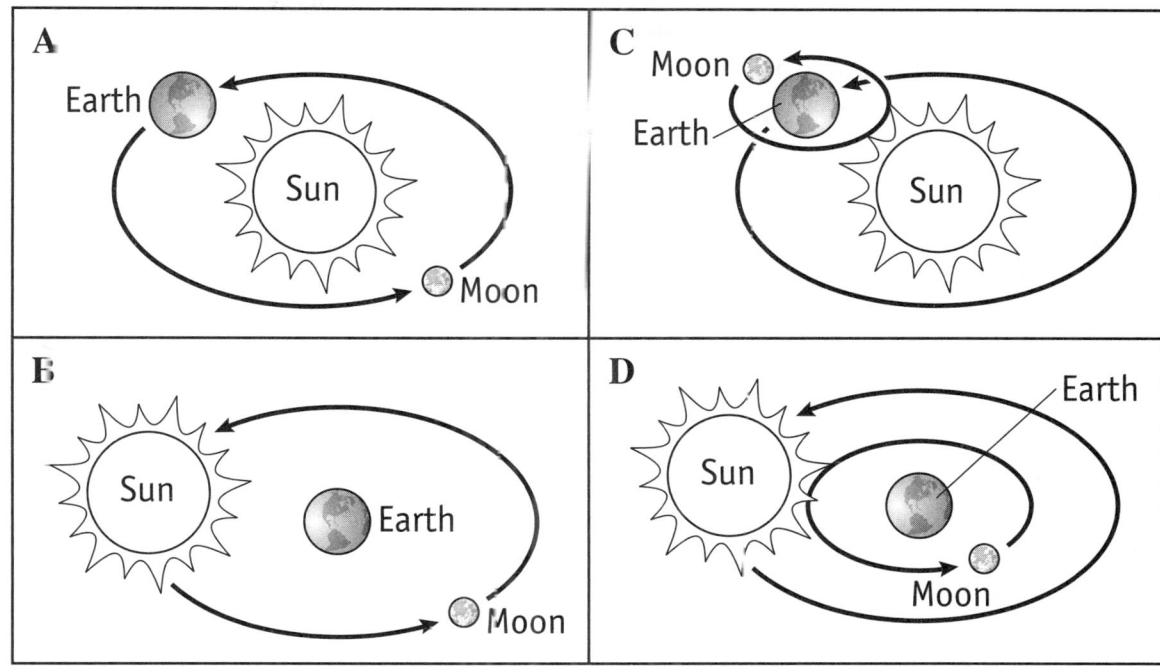

6 This photograph shows sunspots, which are darker than the rest of the sun's surface. What explains why sunspots appear darker?

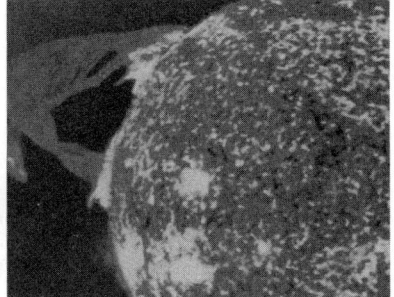

   F They are comets striking the sun's surface.

   G They are cooler areas than the rest of the sun's surface.

OBJ. 4
3.11 (D)

   H They are areas of dark-colored minerals.

   J They are black mountains on the sun's surface.

7 Which of the following best explains why the amount of daylight increases each day in Texas as the summer months approach?

   A The Earth is moving closer to the sun.

   B The Northern Hemisphere is tilting more toward the sun.

OBJ. 4
5.6 (A)

   C The rotation of the Earth is slowing down.

   D The moon is reflecting more light from the sun.

8 What causes the ocean tides to rise and fall on Earth?

   F The magnetic field of the Earth

   G Friction from the ocean currents

OBJ. 4
5.6 (A)

   H Air resistance between wind and water

   J The gravitational pull of the moon

9 The moon appears to change its shape from Earth because the moon —

 A reflects the sun's light

 B acts as a giant lens

 C casts a shadow on the sun's surface

 D has minerals that produce light

OBJ. 4
5.6 (A)

10 Using the information in the table, which planet would you predict revolves around the sun in the shortest period of time?

 F Neptune

 G Jupiter

 H Earth

 J Mercury

OBJ. 4
3.11 (C)

Planet	Distance from Sun (millions of kilometers)
Mercury	57.9
Venus	108.2
Earth	149.7
Mars	227.9
Jupiter	783.1
Saturn	1,427.0
Uranus	2,871.1
Neptune	4,497.2

11 What determines the length of one day on Earth?

 A The time it takes the moon to circle the Earth

 B The time it takes the Earth to circle the sun

 C The time it takes the Earth to spin around on its axis

 D The time it takes the sun to spin around on its axis

OBJ. 4
5.6 (A)

12 A student made a model of the solar system. She used plastic balls to represent the planets. She put the ball for Mercury 10 cm from the sun, and the ball for Saturn 90 cm from the sun. She could make her model more accurate by —

 F placing Mercury farther from the sun

 G placing Saturn closer to the sun

 H switching the balls for Mercury and Saturn

 J moving the ball for Saturn farther from the sun

OBJ. 1
5.3 (C)

13 Maps of the Western Hemisphere and the Eastern Hemisphere are shown to the right. Which two cities have summer at about the same time of year?

OBJ. 4
5.6 (A)

 A Sydney and Tokyo

 B Tokyo and Buenos Aires

 C Buenos Aires and Sydney

 D New York and Buenos Aires

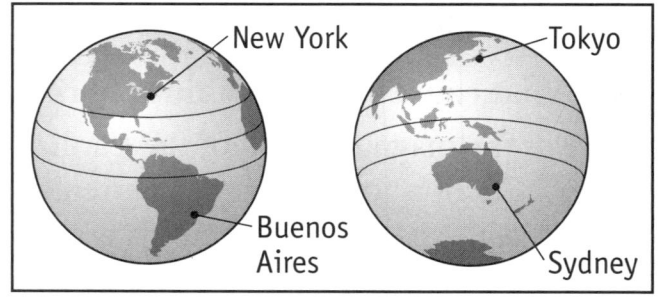

14 **Which of these bodies revolves around a planet?**

F Asteroid   H Comet
G Star   J Moon

OBJ. 4
5.6 (A)

15 **Why is the Northern Hemisphere warmer in summer than in winter?**

A Less sunlight shines on the Northern Hemisphere in summer.
B Earth is moving more quickly in its orbit around the sun.
C The sun gives off more heat in the summer than in the winter.
D More direct sunlight shines on the Northern Hemisphere in summer.

OBJ. 4
5.6 (A)

16 **Which best explains why the sun appears to rise and fall in the sky each day?**

F The Earth rotates.
G The sun rotates.
H The Earth revolves around the sun.
J The sun revolves around the Earth.

OBJ. 4
5.6 (A)

17 **How long does it take for Earth to complete one orbit around the sun?**

A One day   C One year
B One month   D Two years

OBJ. 4
5.6 (A)

18 **If all the planets were to start at the exact same time and circle the sun, which of the following planets would finish the trip first?**

F Mercury   H Jupiter
G Neptune   J Saturn

OBJ. 4
3.11 (C)

19 **The sun was was formed from hot gases and dust in space just like other —**

A planets   C stars
B comets   D moons

OBJ. 4
3.11 (D)

20 **In the illustration below, which of these best represents Neptune?**

F W
G X
H Y
J Z

W   X Earth Y   Z

Sun ○   ○   ○   ○   ○

OBJ. 3
5.7 (A)

21 **Which of the following best describes the surface of the moon?**

    A Moderate temperatures and frequent rainfall

    B Strong winds and warm temperatures

    C Extreme heat in sunlight and cold in shade

    D Frequent sandstorms

OBJ. 4
5.12 (C)

22 **The sun can best be described as a —**

    F lifeless, rocky environment   **H** ball of hot gases

    G planet collapsing into itself   **J** cold, dead star

OBJ. 4
3.11 (D)

23 **The illustration to the right shows Earth tilted on its axis. What is an important effect of this tilt?**

    A Day and night

    B The four seasons

    C The revolution of the Earth

    D The phases of the moon

OBJ. 4
5.6 (A)

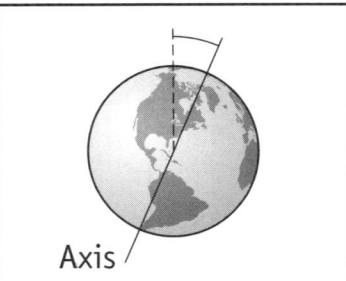

A student used different objects to create a model of the solar system.

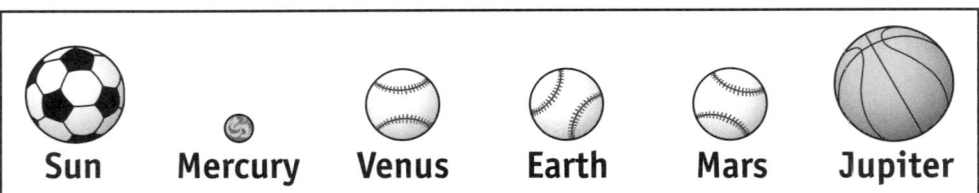

24 **To make the model more accurate, the object representing which of the following should be made larger?**

    F Sun   **H** Venus

    G Mercury   **J** Mars

OBJ. 1
5.3 (C)

Use the diagram below to answer the following question.

OBJ. 4
5.6 (A)

25 **Which of the following is the next phase of the moon?**

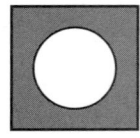      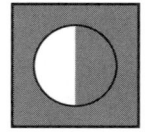   

    A   **B**   **C**   **D**

CHAPTER 10

OUR CHANGING EARTH

In this chapter, you will learn about our planet, the Earth. You will also learn about the materials that make up the Earth, and the processes that help to shape its land forms.

— MAJOR IDEAS —

★ The Earth is made up of different kinds of **materials**, including rocks, soil, water and the gases of the atmosphere.

★ **Soils** have different properties, including texture, the ability to hold water, and the ability to support life.

★ A **process** is a series of events that lead to a change. Different processes have helped shape Earth's surface. Some of these processes have built up the Earth's land forms, including lava flows, sedimentation, and the folding of the Earth's crust.

★ Other processes wear down Earth's surface land forms. These processes include weathering, erosion, earthquakes and the movement of glaciers.

★ Scientists are able to use tree growth rings and the layers of sedimentary rock to draw conclusions about Earth's past.

EARTH'S MATERIALS

The Earth is made up of many different materials. These include rocks, soils, water and gases. These materials are natural resources we all use and enjoy.

ROCKS

The Earth is made of rock, from the tallest mountain to the floor of the deepest ocean. This rock is hard and dense near the surface, but as you go deeper into the Earth the temperature rises and the rock becomes molten.

Rocks are quite familiar to you. You see them all around in the form of mountains, canyons and riverbeds. The outer layer of rock closest to Earth's surface is known as **crust**. A **rock** is any solid found in the Earth's crust or below that is made of one or more minerals. Many minerals form geometric shapes known as **crystals**. Scientists often classify rocks based on their color, size, shape, hardness, mineral crystals, and how the rock was formed.

WATER

Most of the Earth's surface is covered by water. Most of this water is found in the oceans. Ocean water has salt and other minerals dissolved in it. Some water is also frozen in the Earth's ice caps — at the North and South Poles. When it rains and the ground soaks up the rainwater, where does it go? Groundwater is fresh water that collects underground. A very small amount of the Earth's water is found in freshwater rivers, lakes and streams. All living things on Earth require some water to exist.

APPLYING WHAT YOU HAVE LEARNED

DISTRIBUTION OF EARTH'S WATER

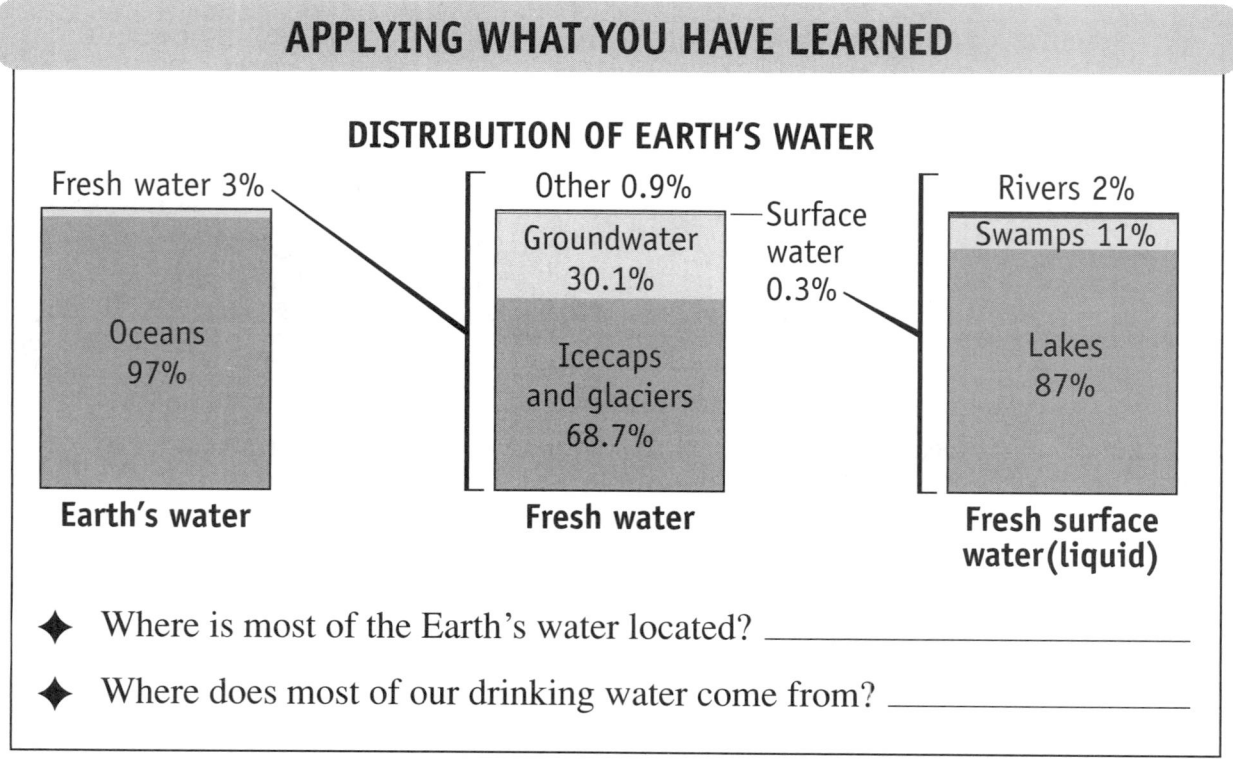

◆ Where is most of the Earth's water located? _____

◆ Where does most of our drinking water come from? _____

SOIL

Some people just call it "dirt," but **soil** is essential to our existence. Soil is necessary for growing crops to feed humans and animals. It is important for plants because the soil stores nutrients and provides support for plant roots and stems.

APPLYING WHAT YOU HAVE LEARNED

✦ What role does soil play in helping to maintain life on Earth? _____

_____

_____

What is Soil? Weathering breaks down rocks on Earth's surface. The material left from the rocks mixes with decaying plants and animals to make soil. It usually takes hundreds of years to form one inch of soil. **Soil** is therefore a mixture of many materials including sand, clay, rock, water, fungi, bacteria, and decayed plants and animal material (*humus*). There are different types of soil based on the mix of materials found in each type. Each type of soil has its own texture or feel, its own ability to hold water, and its own ability to support life. Soil is able to store water and nutrients used by plants.

Types of Soil. Although there are many ways to describe different types of soil, most scientists rely on how much sand, silt, and clay there is in the soil. Sand is a small stone particle in the soil. Silt feels smooth and powdery, while clay is the smallest type of particle found in soil. Clay and dead plant and animal material can hold water. Soils with a large amount of clay and decayed material will hold more water than sandy soils.

If a particle of **sand** were the size of a

basketball,

then **silt** would be the size of a

baseball,

and **clay** would be the size of a

golf ball.

Soil texture is based on how large the pieces of clay and other particles in the soil are, and how much decayed plant and animal life there is. Soils also contain different chemicals, like salts. These chemicals affect the ability of soil to support life. It is possible to change the soil by adding different materials and chemicals. Farmers may add more dead plant and animal material (*humus*) or special chemicals to the soil to help crops grow.

THE GASES OF THE ATMOSPHERE

Our Earth is surrounded by a blanket of air that we call the **atmosphere**. This atmosphere reaches almost 350 miles above the surface of Earth. Living at the Earth's surface, we are only able to see what occurs in the atmosphere closest to the ground.

Our atmosphere is a mixture of different gases. Most of the atmosphere is made up of nitrogen. Both humans and animals need oxygen to live, which they get from the Earth's atmosphere. Plants take carbon dioxide from the atmosphere, which they need for photosynthesis.

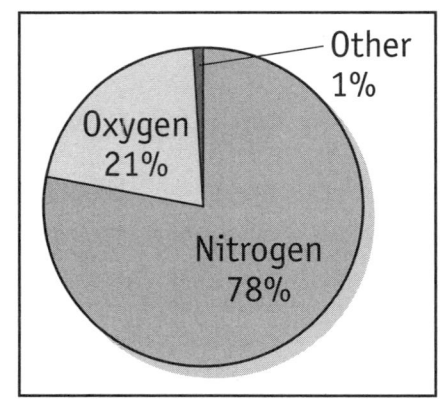

Why Do We Need the Atmosphere? Life on Earth is supported by the atmosphere. The gases of the atmosphere absorb energy from the sun. The atmosphere also plays a role in recycling water and other chemicals. It provides a moderate climate for us to live in. The atmosphere protects living things from the harmful effects of ultraviolet radiation given off by the sun, and slows down meteors that might strike the Earth.

APPLYING WHAT YOU HAVE LEARNED

Complete the following chart.

EARTH'S MATERIALS

Name	Where Found	Characteristics
Rocks		
Soil		
Water		
Gases		

FORCES SHAPING EARTH'S SURFACE

Different processes have helped shape Earth's different land forms.

CONSTRUCTIVE FORCES

Some processes help to build up the Earth's land forms. These processes are known as **constructive forces**. They take long periods of time to have measurable effects.

FOLDING TO BUILD MOUNTAINS

Sometimes parts of Earth's crust squeeze together. When this happens, over millions of years, the Earth's crust may start to fold. This folding pushes some of Earth's crust upwards. This process can form **mountains**. Many of the great mountain ranges on Earth were created by this very slow movement of the Earth's crust.

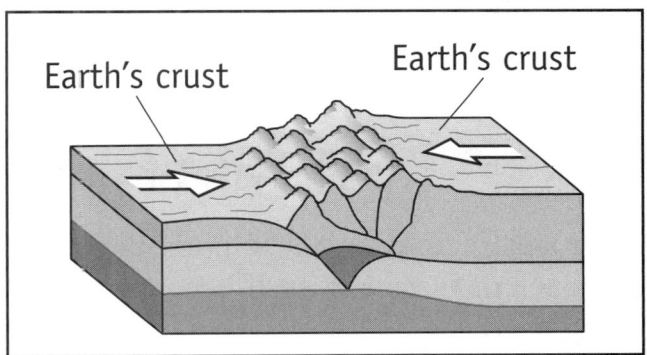

VOLCANOES AND LAVA FLOWS

A **volcano** is an opening in the Earth's surface that lets out molten rock and gases. Volcanoes often occur at the edges of huge plates that make up the surface of the Earth.

Sometimes, part of a plate will sink into the Earth, where it creates molten rock known as **magma**. The magma escapes through a hole in the Earth's crust. Once this molten rock comes to the surface, it is known as **lava**. As lava builds up on the ground around the volcano, it often gives the volcano a typical cone shape.

Mt. St. Helens in Washington State erupts.

Many islands and mountains have been formed by volcanoes. For example, the Hawaiian Islands are actually the tops of volcanoes in the Pacific Ocean. Lava may also come out of cracks in the Earth's surface from an earthquake. A **lava flow** will leave a smooth, flat, rock-hard surface. Cracks appear when the rock shrinks as it cools. Often, several layers of lava flows build up on top of each other.

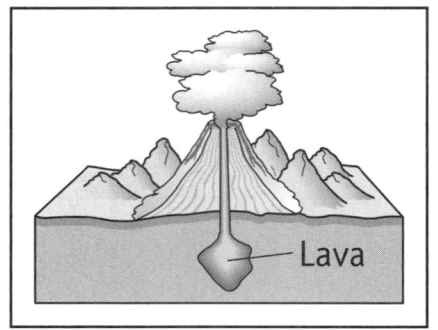

Lava pushes up from
beneath the Earth's surface

SEDIMENTATION

Sedimentary rocks are made by sand, mud and small pieces of rock or shells that are picked up by wind, water or ice. These pieces of rock often are moved to other locations or collect at the bottom of lakes and oceans. They get pressed together by the weight above. Over time, these layers of sand and mud, often at the bottom of lakes and oceans, turn to rock. Scientists call these sedimentary rocks. One characteristic of sedimentary rock is that it has layers.

APPLYING WHAT YOU HAVE LEARNED

Constructive forces help to build up the Earth's land forms. Explain how each of these forces affects the Earth's surface.

Constructive Force	How it Helps Build up Land Forms
Folding of the crust	
Volcanoes and lava flows	
Sedimentation	

DESTRUCTIVE FORCES

The Earth's surface is constantly undergoing changes. Just as some processes help to build up the Earth's land forms, other forces are at work slowly tearing them down.

WEATHERING

The wearing down of rocks on the Earth's surface by the actions of wind, water, ice and living things is referred to as **weathering**. Water, for example, expands when it freezes. Cool nights and hot days often cause rocks to crack and break apart. Water may seep into cracks in rocks and expand these cracks if the water freezes. Rain and running water will also break down rock

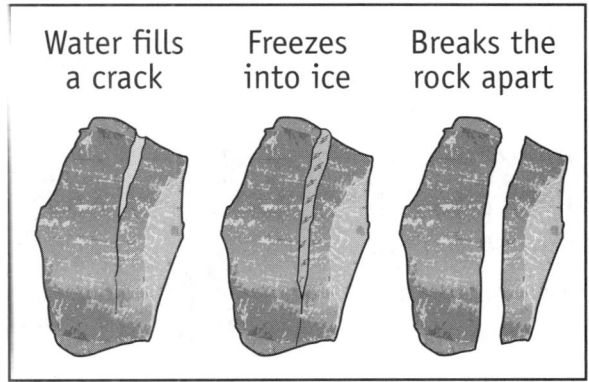

into smaller particles. Running water also smoothes rock, creating round stones and pebbles. Some chemicals **dissolve** rocks. Plants wedge their roots into the cracks of rocks, spreading them apart. Microscopic organisms may also cause rocks to break down and disintegrate.

APPLYING WHAT YOU HAVE LEARNED

Describe some of the different kinds of weathering that wears down rocks:

_____

_____

EROSION

The process by which soil and rock are broken down and moved away is known as **erosion**. Once rock is broken into smaller particles, the wind, running water, ice or gravity may cause these particles to move to a new location. If you've ever been to a beach on a windy day, you can understand the power of sand blown by the wind. Rivers also carry and deposit sediment elsewhere. The action of ocean waves can wear down a rocky shoreline or move the

Erosion at the Grand Canyon (USGS)

sand on a beach into the ocean, causing beach erosion.

GLACIERS

Glaciers are rivers of ice. They are formed in areas where there are very cold winters and cool summers. The snow that falls in the win-

ter does not melt during the summer. Instead, it turns to ice. New snow then falls on top of this ice. As the layers of snow build up, the weight of the snow increases. This weight pushes on the ice and snow below, creating very thick, dense sheets of ice called **glaciers**.

Glaciers actually move quite slowly. As they move, they scrape the Earth's surface, picking up loose rock in their way, digging holes, wearing down mountains, and moving rocks and soil. The rocks and boulders carried by the glaciers scrape the surface. A glacier can move millions of tons of material. Often, moving glaciers will carve valleys through mountains. When it stops, a glacier will often leave behind rich soil. During the last Ice Age, glaciers covered much of what is now the United States.

APPLYING WHAT YOU HAVE LEARNED

◆ What is a glacier? _____

_____

_____

◆ How are glaciers formed? _____

_____

_____

◆ How do glaciers cause erosion? _____

_____

_____

EARTHQUAKES

Sometimes movements of the Earth's crust, as pieces of crust slide against each other, create pressure and stress in the surrounding rock. Eventually, the rocks release the energy created by this stress in an **earthquake**. The rock vibrates to release this stress. This energy passes up through the Earth to the surface in a series of seismic waves. An earthquake may tear down an

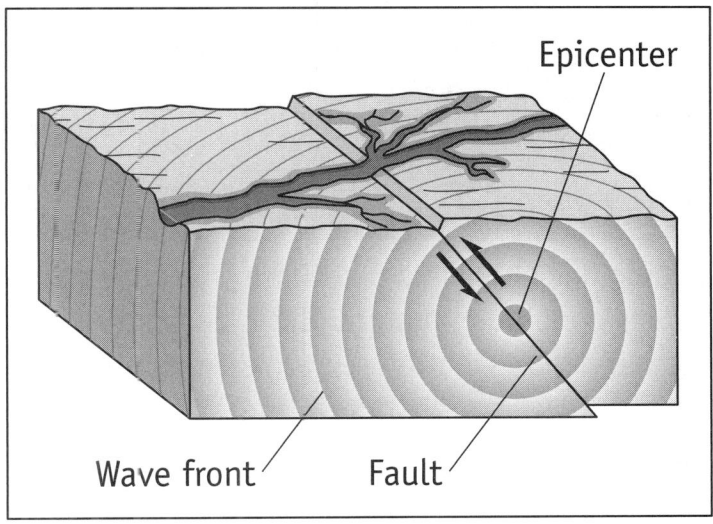

area or help build it up. Lava may come through cracks created by the earthquake, creating a mountain range. The earthquake may shift the Earth's crust, lowering some areas, while raising others up.

APPLYING WHAT YOU HAVE LEARNED

Destructive forces help to wear away the Earth's land forms. Explain how each of these processes affect the Earth's surface.

Destructive Force	What is it?	Effects on Earth's Surface
Weathering		
Erosion		
Glaciers		
Earthquakes		

EARTH'S PAST: DRAWING CONCLUSIONS ABOUT "WHAT HAPPENED BEFORE"

People once believed that Earth was only a few thousand years old. In the 1830s, scientists concluded that the Earth is actually much older than that. The layers of sedimentary rock we now see took millions of years to form. There are many different ways scientists now determine what events happened on Earth thousands or even millions of years ago. Two of these ways are by studying tree-ring growth and the layers of sedimentary rock.

TREE-GROWTH RINGS

Every tree keeps its own unique "diary" of changes in climate and other events that affected its growth. Each year a new page is added to the tree's diary. Each year, beneath its bark, a tree adds a new layer of wood to its trunk. This makes the tree wider in

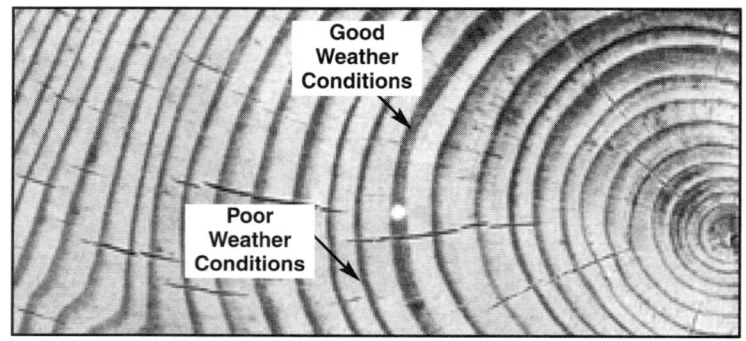

diameter. If the tree is cut down, its layers appear as a series of rings, one inside the other. These are called **tree-growth rings**.

Scientists can tell the age of a tree by counting its rings. Some trees are hundreds of years old, and a few are even more than a thousand. Scientists can also tell about an area's past. When weather conditions were good, the ring for that year will be thick. If the weather is too dry, if insects destroy most of the tree's leaves, or if there is some other trouble, the tree-growth ring will be thin.

APPLYING WHAT YOU HAVE LEARNED

◆ What can studying a tree's rings tell scientists? _____

_____

_____

SEDIMENTARY ROCKS

Scientists are also able to tell about an area's past by looking at its **sedimentary rocks**. Sedimentary rock, you may recall, is formed by sand, mud, and other particles falling to the ground or to the floor of a lake or ocean.

Scientists who study the past try to put things in time order. They assume that the oldest layers of rock are on the bottom and that the most recent layers appear on top. Think about newspapers piling up in a recycling bin. Monday's newspaper goes in first, followed by Tuesday's on top of it. Then Wednesday's newspaper goes on top of that. The top layer of newspapers — Wednesday's — is the most recent.

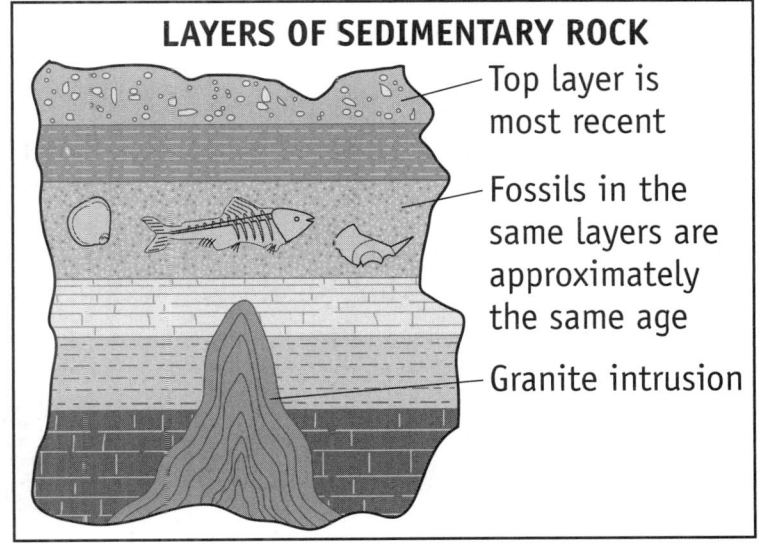

LAYERS OF SEDIMENTARY ROCK

Top layer is most recent

Fossils in the same layers are approximately the same age

Granite intrusion

It is the same with sedimentary rocks. The most recent rocks are on top. Scientists use this knowledge to estimate the age of rocks relative to each other. Scientists are also often able to use evidence from a rock layer to determine the type of environment that existed when the layer was first formed. For example, limestone is formed from the shells of sea creatures. Whenever limestone is found, scientists assume there was once a marine environment.

Fossils are impressions created by the remains of dead plants and animals in sedimentary rocks. Sometimes, a dead plant or animal will leave behind bones, shells, leaves, or tracks of a past life on Earth. For example, a dinosaur walks on mud, leaving its footprint. The mud dries, and sand settles on the mud footprints. The sand and mud harden into different types of

Fossil of a fish embedded in a rock

sedimentary rock. The dinosaur's footprint can be seen when the rock is dug up. By examining fossils, scientists can often estimate the age of a sedimentary rock. If they recognize the plant or animal that created the fossil, they can guess the rock's age.

WHAT YOU SHOULD KNOW

A. You should know that the Earth is made up of different kinds of **materials**, including rocks, soil, water and the gases of the atmosphere.

B. You should know that **soils** have different properties, including texture, the ability to hold water, and the ability to support life.

C. You should know that a **process** is a series of events that lead to a change. Different processes have helped shape Earth's surface. Some have built up the Earth's land forms, while others wear down Earth's surface land forms.

D. You should know that scientists are able to use **tree-growth rings** and the **layers of sedimentary rock** to draw conclusions about Earth's past.

CHAPTER STUDY CARDS

Earth's Materials

Earth has many different types of materials:

★ **Rock**. Any solid found on Earth's crust made of minerals.

★ **Soil**. Material from ground up rock and decayed plants and animals.

★ **Water**. Most of Earth's surface is covered by water: saltwater, freshwater and groundwater.

★ **Gases**. Earth's atmosphere is a mixture of different gases; mainly nitrogen and oxygen.

Constructive Forces

Constructive forces act to build up Earth's land forms. They consist of:

★ **Folding of Earth's Crust**. Helps to build up mountains and hills.

★ **Volcanoes and Lava Flows**. Molten rock from underground breaks through Earth's surface and hardens when cool.

★ **Sedimentation**. Sand, mud, small pieces of rock and shells are deposited and pressed together into sedimentary rock.

Destructive Forces

Destructive forces act to wear down the Earth's land forms. They consist of:

★ **Weathering**. When wind, water, ice or living things wear down rocks.

★ **Erosion**. When soil and rock are broken down and carried away by wind and water.

★ **Glaciers**. Giant sheets of ice that move slowly, scraping the Earth's surface.

★ **Earthquakes**. Vibrations of Earth's crust.

Clues to Earth's Past

Scientists use different ways to tell what happened in an area in the distant past.

★ **Tree-Growth Rings**. Each year, trees grow one new layer or ring. In wet weather a tree ring is usually thicker; in dry weather, the ring is much narrower.

★ **Sedimentary Rock**. Scientists examine its layers to see what an area was once like. Rock made from shells shows it was once under water. **Fossils** help tell the rock's age.

CHECKING YOUR UNDERSTANDING

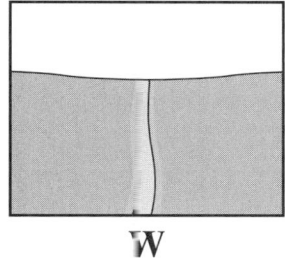 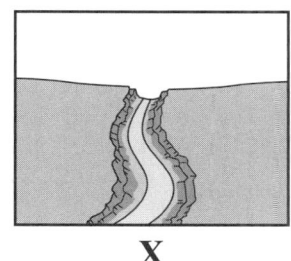 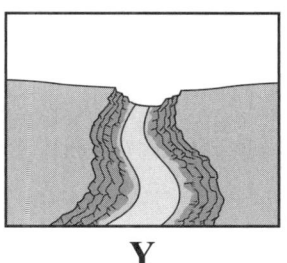 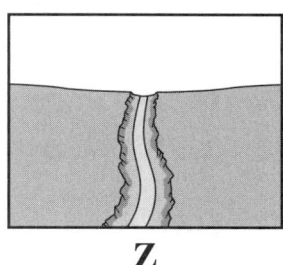

W   X   Y   Z

1   These diagrams show how a river can cause the erosion of land. Which is the correct order of letters that show how the river has aged from youngest to oldest?

A   Z → W → X → Y
B   W → Z → X → Y
C   X → W → T → Z
D   Y → W → X → Z

OBJ. 4
5.12 (A)

HINT

To answer this question, you need to know how water erosion gradually grinds down rock and moves rock particles away. As the river continues to flow, it makes the gap wider. The correct answer goes from the smallest (W) to the largest (Y) — **Choice B**.

Now try answering some additional questions on your own:

2   When two pieces of crust slide against each other, what is most likely to occur?

F   Tornado
G   Earthquake
H   Hurricane
J   Water erosion

OBJ. 4
3.6 (B)

3   A farmer sees an area that will be good for farming because of its rich topsoil. Which of these processes contributed most to this rich top soil?

A   Ocean currents
B   Lava flows
C   Folding of Earth's crust
D   Decaying of plant life

◆ Examine the Question
◆ Recall What You Know
◆ Apply What You Know

OBJ. 4
4.11 (A)

The illustration below shows erosion caused by ocean waves.

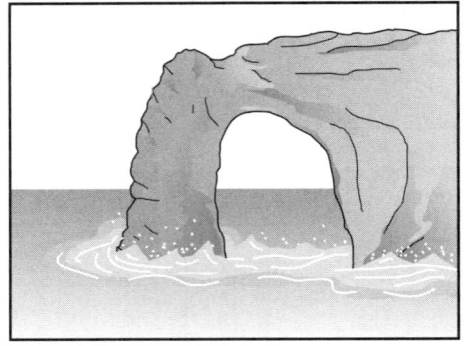

4 What will most likely happen to the hole in this landform as erosion from ocean waves continues?

   F It will be buried by sand.

   G It will be covered by ocean water.

   H It will become larger.

   J It will remain the same.

OBJ. 4
5.12 (A)

5 How do some plants break large rocks into smaller pieces?

   A Plant roots grow into cracks in rocks, forcing them apart.

   B Plant stems surround and squeeze rocks until they break.

   C Plant leaves insulate rocks from extreme temperatures.

   D Seeds from plants fall into rocks and release powerful chemicals.

OBJ. 4
5.12 (A)

6 Why are most fossils found in sedimentary rocks?

   F Sedimentary rocks are not very old.

   G Sedimentary rocks are found only under the oceans.

   H Organisms live only in areas where sedimentary rock is found.

   J Sediment covers where a plant or animal left an impression or skeleton.

OBJ. 4
5.11 (B)

7 A student mixes together sand, clay, and decayed material from dead plants. This mixture would be most useful for learning about —

   A rocks

   B gases

   C soils

   D fossils

◆ Examine the Question
◆ Recall What You Know
◆ Apply What You Know

OBJ. 4
4.11 (A)

8 Scientists study tree rings to learn about an area's past. What would a scientist conclude about one year in an area's past after examining a tree-growth ring that was thicker than other rings?

   F Weather conditions were good.

   G The tree's leaves fell off early.

   H The area faced insect attacks.

   J There was a severe drought.

OBJ. 4
5.11 (B)

9   A science class built the models shown to the right to conduct an experiment. Students poured the same amount of water over both models. They observed that in Model A, most of the dirt emptied out of the container. In Model B, only a small amount of dirt emptied

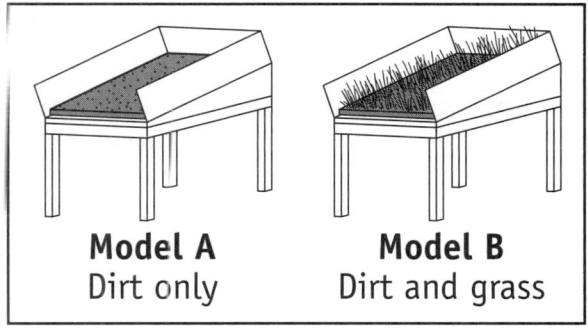

Model A
Dirt only

Model B
Dirt and grass

out of the container. What were the students studying in their experiment?

   A   The impact of volcanoes
   B   The force of earthquakes
   C   The rock cycle
   D   The effects of water erosion

OBJ. 4
4.11 (A)

10   The most common cause of earthquakes is —

   F   the sinking of the ocean floor
   G   unequal heating of the atmosphere
   H   giant waves caused by the pull of the moon
   J   movements of the Earth's crust

OBJ. 4
3.6 (C)

11   How does freezing water cause rocks to weather?

   A   It holds them in place.
   B   It makes them longer.
   C   It makes them thicker.
   D   It makes them crack.

OBJ. 4
5.12 (A)

12   This rock was brought to school. The class found fossils of water plants and shells in the rock. What does this tell us about the rock?

Fossil

Fossil

   F   The rock is gray and brown.
   G   The rock must be washed off first.
   H   The rock is heavier than most rocks in the area.
   J   The rock was once at the bottom of a lake or sea.

OBJ. 4
5.11 (B)

13   What most likely causes rocks in a stream to be smooth?

   A   Movement of fish
   B   Air currents
   C   Earthquakes
   D   Erosion

♦ Examine the Question
♦ Recall What You Know
♦ Apply What You Know

OBJ. 4
5.11 (A)

14 A student conducts an experiment. She puts different types of soil in four identical pots. Each pot has a hole in the bottom. She puts the same amount of water in each pot. A different amount of water drains out of each pot. Which would be the best conclusion to draw from this experiment?

F Different soils have different colors.

G Different soils have different textures.

H Different soils have different nutrients to support life.

J Different soils have different abilities to absorb water.

OBJ. 4
4.11 (A)

15 Wave action against solid rock can cause changes in the structure of a rock. What is the correct sequence of erosion of the rock formation in the diagrams?

A $1 \rightarrow 2 \rightarrow 3 \rightarrow 4$

B $3 \rightarrow 4 \rightarrow 2 \rightarrow 1$

C $4 \rightarrow 2 \rightarrow 3 \rightarrow 1$

D $2 \rightarrow 3 \rightarrow 4 \rightarrow 1$

OBJ. 4
5.12 (A)

◆ Examine the Question
◆ Recall What You Know
◆ Apply What You Know

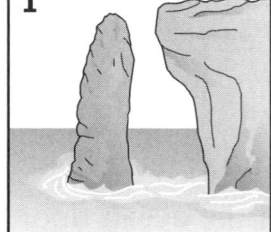
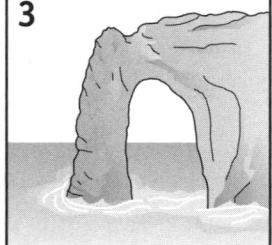
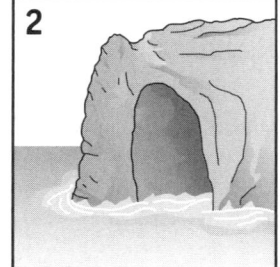
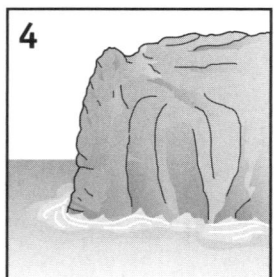

16 The map shows the location of Greenland. Today Greenland has a cold climate with few plants. However, fossils of many plants have been found in Greenland. Which best explains why plant fossils were found there?

F Greenland has a short winter.

G Greenland once had a warm climate.

H Most plants are able to survive in cold weather.

J Plants were brought there from places with tropical climates.

OBJ. 4
5.11 (B)

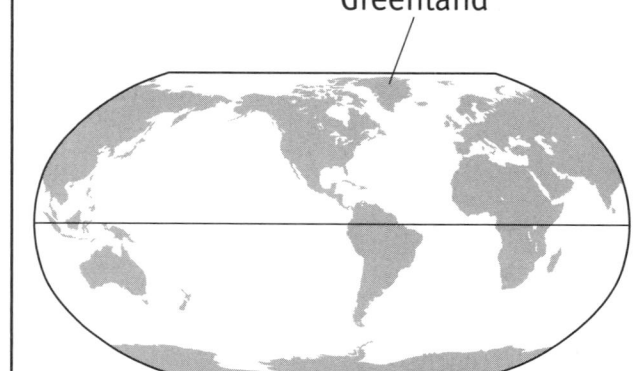

17 Rocks that formed in areas that were once covered by oceans often contain fossils of animals that lived in the sea. Which of these rocks were once covered by ocean waters?

OBJ. 4
5.11 (B)

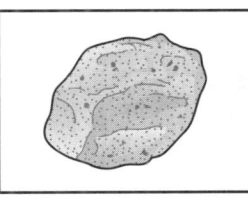
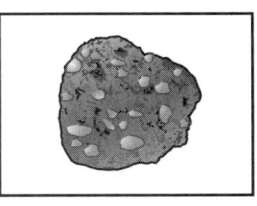

A   B   C   D

SOIL AND WATER ABSORPTION RATES

Soil Type	Time for 1 Inch of Water to Drip Through
Sand	0.5 hours
Loam	2.0 hours
Silt loam	2.25 hours
Clay	5.0 hours

18 A scientist tested four types of soils to find which type was best at holding water. The results of the investigation are recorded in the table above. According to this information, which soil type was best at retaining water?

F   Sand
G   Loam

H   Silt loam
J   Clay

OBJ. 4
4.11 (A)

19 This picture shows the tree rings of a tree found in many regions of the United States. In studying these growth rings, what might scientists conclude about weather conditions in Year A and Year B?

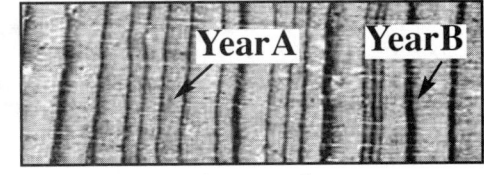
Growth-rings of a tree

A   Rainfall was similar in both years.
B   Rainfall was low in both years.
C   Rainfall was better in Year A than B.
D   Rainfall was better in Year B than A.

OBJ. 4
5.11 (B)

20 The surface of the Earth can be changed by the force of a moving glacier. Which change to the land would most likely be caused by the force of a moving glacier?

OBJ. 4
3.6 (B)

F   Building up mountains
G   Destroying rich soil

H   Bringing a lava-flow
J   Cutting a valley through mountains

CHAPTER 11

EARTH'S RESOURCES, INTERACTIONS AND CYCLES

In this chapter, you will learn about Earth's natural resources; about its oceans, atmosphere and land forms; and about its water, carbon and nitrogen cycles.

— MAJOR IDEAS —

★ The Earth has **renewable**, **nonrenewable**, and **inexhaustible resources**.

★ **Past events** have led to the formation of many of the Earth's resources.

★ The Earth's different systems **interact** with each other. The Earth's oceans, atmosphere and land forms all influence one another.

★ Earth's **water**, **carbon** and **nitrogen** are continually recycled.

★ Because of its atmosphere, soil, oceans, and living things, Earth's physical characteristics are different from those of the moon.

TYPES OF NATURAL RESOURCES

A **natural resource** is something found in nature that people are able to use to meet their needs. All of the Earth's resources can be classified into three types:

Renewable Resources   Nonrenewable Resources   Inexhaustible Resources

RENEWABLE RESOURCES

A **renewable resource** is something that can be replaced, such as wood from trees. A renewable resouce can be replaced by natural processes like growth.

158

For example, wood can be replaced because new trees can be regrown. If a tree is planted each time one is cut down for lumber, the resource will eventually grow back and be renewed. However, enough time must be allowed for the resource to renew itself. Forests, livestock, and food crops are all considered renewable resources.

Forests are renewable resources.

NONRENEWABLE RESOURCES

A **nonrenewable resource** is formed over a very long period of time and cannot be replaced or renewed. For example, oil, coal, copper and other minerals are nonrenewable resources. Once used, they cannot be replaced.

INEXHAUSTIBLE RESOURCES

An **inexhaustible resource** is in such large supply that it cannot be used up by human activity. Its supply cannot be exhausted. Energy from the sun is an example of an inexhaustible resource. Heat from deep inside the Earth and wind power are also inexhaustible resources. People cannot use them up.

APPLYING WHAT YOU HAVE LEARNED

Classify each of the following resources by checking the correct box:

	Renewable	Nonrenewable	Inexhaustible
Gold	☐	☐	☐
Iron ore	☐	☐	☐
Wind energy	☐	☐	☐
Nitrogen gas	☐	☐	☐
Coal	☐	☐	☐
Fresh water	☐	☐	☐
Oak trees	☐	☐	☐
Rubber plants	☐	☐	☐
Wild Salmon	☐	☐	☐

USING EARTH'S MATERIALS AS RESOURCES

Each of Earth's different types of materials, which you learned about in the last chapter, can be used as a **resource**. Let's determine which of these resources would be considered as **renewable**, **nonrenewable** or **inexhaustible**.

SOIL

Soil is a very important resource for growing crops. Some soils can grow different crops better than others. Because of this, each type of soil is a **renewable resource**. For example, if one type of soil has a certain kind of decayed plant life in it, more of it can be made by adding more of the same decayed plant life to the clays and sands already in the soil.

To preserve the soil against erosion, farmers must take measures of **soil conservation**. These include planting crops that hold the soil together against the wind. Plants also slow down water as it flows over the land, and they allow rainwater to soak into the ground. Plants hold the soil in place and prevent it from being washed or blown away. Farmers can also plow across instead of up and down on hills and slopes to reduce soil erosion from running water.

Soil erosion from water flow.

APPLYING WHAT YOU HAVE LEARNED

◆ List the factors that cause **soil erosion** and some **measures** that farmers can take to limit this erosion. _____

_____

_____

_____

GASES

Gases in the atmosphere provide another important resource. Since most of the atmosphere is **nitrogen**, this gas is almost inexhaustible. However, as companies and people pollute the air, its nitrogen might become unusable. **Oxygen** is a very important **renewable resource**. We cannot live without oxygen to breathe. Oxygen is produced by plants during photosynthesis. Scientists believe the Earth did not always have oxygen in its atmosphere. Gradually, over millions of years, plants and other living things (*special bacteria*) produced the oxygen in our atmosphere. If we reduce plant life by cutting down rainforests, we may reduce our oxygen.

A forest in California destroyed by too much pollution in the air.

The breathing of animals and the burning of materials create a gas called **carbon dioxide**. The burning of gasoline in car engines and of oil to heat homes and run factories creates huge amounts of carbon dioxide. Too much carbon dioxide can hurt our environment. At one time, human activity released very few gases into the air. Today, population growth, fossil fuel burning and deforestation are changing the mixture of gases in the atmosphere. Some scientists think the large amount of carbon dioxide now being produced has caused a heating up of Earth's surface, known as **global warming**.

APPLYING WHAT YOU HAVE LEARNED

◆ Why are scientists so concerned about the future balance of oxygen and carbon dioxide in the atmosphere? _____

_____

_____

WATER

Water is needed by all living things to survive. Some plants and animals live in the Earth's oceans. All other plants and animals require fresh water. People also use fresh water to manufacture many products. Fresh water is a very valuable **renewable resource**. Its supplies are constantly being rebuilt because of the **water cycle** (*see page 166*). Water pollution threatens this resource.

ROCKS

Rocks above and below the Earth's surface contain many valuable minerals and other resources. Some types of rock, like sandstone, are so common they are almost inexhaustible. Silicon comes from sand and rocks. It is so common in rock, it might be considered inexhaustible. It is used to make silicon chips for computers. Minerals like gold and iron ore are more limited. Because rocks and minerals cannot be replaced after they are taken from the ground, they are **nonrenewable resources**.

FOSSIL FUELS

Fossil fuels like coal, oil, and natural gas are very special resources. They can be burned to release large amounts of energy. We burn fossil fuels to run our car engines, heat our homes, power our machinery, and create electricity. Fossil fuels actually come from the remains of ancient living things.

★ **Coal** is a brown or black rock formed from plants in forests and swamps some 400 million years ago. The plants captured energy from sunlight through photosynthesis. After the plants died, they decayed. As the Earth changed, this material became buried. Over millions of years, heat and pressure changed it into coal. Today, we burn coal for electricity and heat. When burned, it releases the energy stored by plants from the sun millions of years ago.

★ **Oil and Natural Gas** are also fossil fuels. They were formed by very tiny one-celled plants and animals in the ocean. They also stored energy from the sun through photosynthesis. When they died, they fell to the ocean floor, where mud and sediment covered them. Over millions of years, heat and pressure changed their soft bodies into liquid oil and natural gas.

OIL (PETROLEUM) AND NATURAL GAS FORMATION

300–400 million years ago

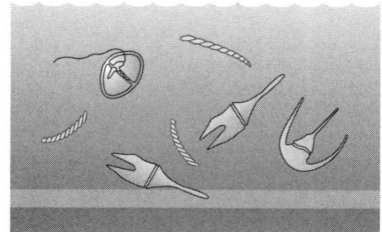

Tiny sea plants and animals died and were buried on the ocean floor. Over time, they were covered by layers of silt and sand.

50–100 million years ago

Over millions of years, the remains were buried deeper and deeper. The enormous heat and pressure turned them into oil and gas.

Today, we drill down through layers of sand, silt, and rock to reach the rock formations that contain oil and gas deposits.

It takes millions of years for fossil fuels like coal and oil to form. They can only be burned once. For this reason, they are important **nonrenewable** resources. As oil, coal and gas are burned, some scientists fear their supply on Earth is being used at a faster rate than new energy sources can be found.

APPLYING WHAT YOU HAVE LEARNED

◆ Why are scientists concerned about the rate humans are using up fossil fuels? _____

_____

◆ What steps can be taken to help the world's supply of fossil fuels last longer? _____

_____

INTERACTIONS OF EARTH'S SYSTEMS

The Earth's land forms, oceans, and atmosphere interact with each other. This interaction has important effects, such as typical weather patterns.

INTERACTION OF OCEANS AND LAND FORMS

One example of this interaction is in the relationship between rivers, oceans, and land forms. Rivers carry sediment from the land into the ocean. Most of the ocean floor is covered by this sediment, which has taken millions of years to accumulate. The ocean also affects the land. Ocean currents carry some of this sediment to coastlines, where it forms sandy beaches. The tide comes twice each day as sea levels rise and fall. At high tide, the sea rises higher on the beach. Tides and waves can erode shorelines, wearing away at rock and dissolving minerals.

The action of tides on a shoreline can lead to serious erosion.

APPLYING WHAT YOU HAVE LEARNED

✦ How do rivers affect the ocean? _____

_____

_____

✦ What impact do tides have on land forms? _____

_____

_____

WEATHER PATTERNS

The **weather** refers to conditions in the atmosphere at the Earth's surface — including temperature, rainfall or snowfall, and wind. **Climate** is the typical weather of a place over a long period of time. For example, Alaska has a colder climate than Texas. Weather results from the interaction of several systems — land features, energy from the sun heating the Earth's atmosphere, the spinning of the Earth creating winds, and the Earth's oceans. For example:

★ Temperatures are usually warmer the closer one gets to the equator — the imaginary line around the Earth's middle. Temperatures become cooler as one goes higher, such as on mountains.

★ Because air cools as it rises over a mountain, the ocean side of a mountain often has heavy rainfall. The air loses its moisture and becomes drier

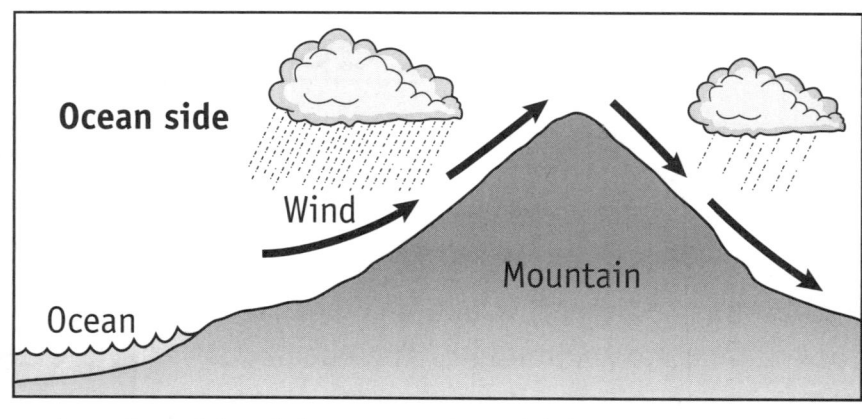

by the time it reaches the other side of the mountain, which has less rain.

★ Differences in the temperature of land and ocean also affect climate. Water requires more energy than land to change its temperature. As a result, oceans and lakes stay cooler than land in summer and warmer than land in winter. This affects air flowing over these areas.

WINDS AND TORNADOES

The spinning of the Earth and the uneven heating of the atmosphere by the sun create typical wind patterns. Cold air sinks and hot air rises. **Tornadoes** are high-speed winds that whirl in a funnel. They often occur in the flat plains of the central United States. A tornado occurs when dry, cool air meets warm, humid air. The warm air rises quickly, sucking in both air and objects.

TROPICAL HURRICANES

Hurricanes occur in tropical regions in late summer and early fall when the ocean water is very warm. The warm ocean water evaporates quickly and rises. Air around the rising air column begins to spin at high speeds. The hot air rises until it cools. Then it releases energy and causes heavy rains, winds and lightning.

APPLYING WHAT YOU HAVE LEARNED

◆ Give two examples of typical weather patterns. _____

_____

_____

_____

◆ How does the sun's energy help to cause weather patterns? _____

_____

_____

THE EARTH'S CYCLES

Many of the processes of Earth's surface are **cycles** — processes that go through a series of steps in which the last step leads back to the first step. Then the process begins all over again. Once a scientist knows the steps of a cycle, he or she can often predict what will happen next in the cycle.

THE WATER CYCLE

Most of Earth's surface is covered by water. The amount of water on Earth has remained fairly constant over time. The glass of water you drank yesterday may have fallen as rain last year. It may have also been used by dinosaurs millions of years ago. This is possible because the Earth's water is in constant movement. The **water cycle** is the process by which Earth's water moves into and out of the atmosphere. The water cycle begins when energy from the sun heats the surface of lakes and oceans. This solar energy causes some of this water to turn into a gas (**water vapor**), which **evaporates** into the atmosphere.

In the Earth's atmosphere, the water vapor **condenses**, or turns back into tiny droplets of liquid water (*condensation*). These droplets form clouds. When the droplets get larger, they fall back to the ground as rain or snow (*precipitation*). Gravity pulls them back to the Earth's surface. Some of this water falls on land where it forms lakes, streams and rivers. Some of the water absorbed by the ground sinks until it hits dense rock and collects as **groundwater**. Some of this water evaporates, but much of it eventually drains back into the ocean. Then the process begins all over again as surface water from the ocean evaporates into the atmosphere.

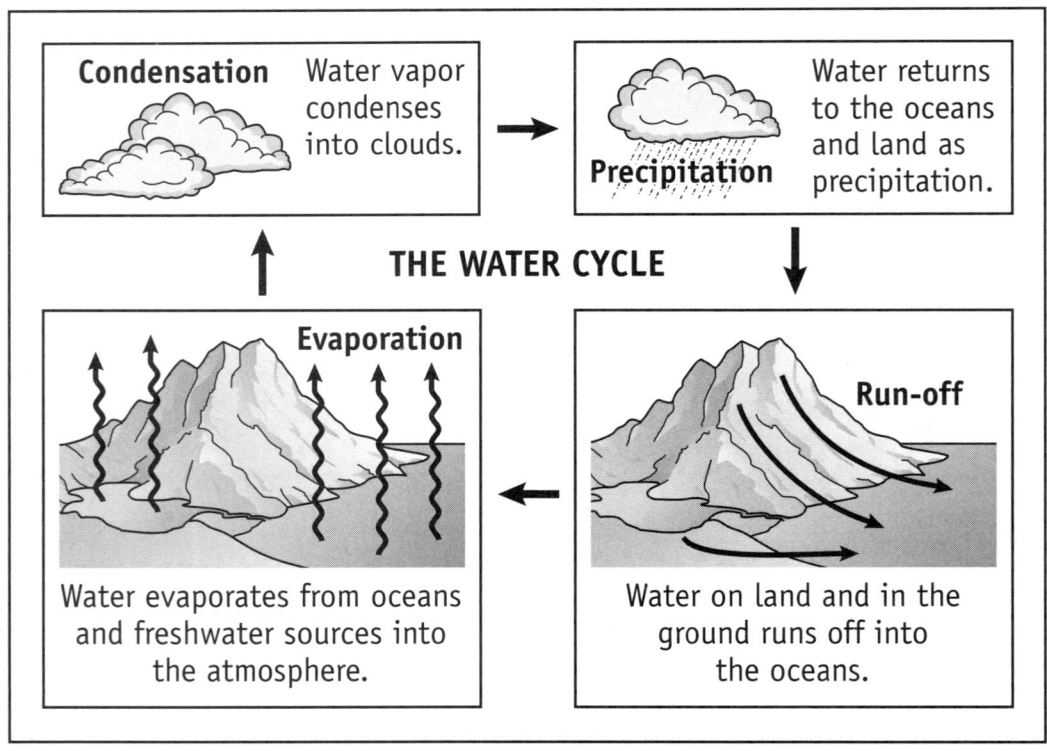

Condensation Water vapor condenses into clouds.

Precipitation Water returns to the oceans and land as precipitation.

THE WATER CYCLE

Evaporation Water evaporates from oceans and freshwater sources into the atmosphere.

Run-off Water on land and in the ground runs off into the oceans.

APPLYING WHAT YOU HAVE LEARNED

◆ List the four main steps of the water cycle. _____

_____

_____

_____

_____

◆ Describe how the water cycle affects the Earth. _____

_____

THE CARBON CYCLE

Carbon is a chemical that is an essential part of life on Earth. It is found in all living things. Carbon is in the food you eat, the clothes you wear, the shampoo you use and the gasoline that fuels your family car. It is the sixth most common chemical in the universe. Carbon is present in the atmosphere, in layers of limestone on the ocean floor, and in fossil fuels. Carbon is continuously recycled among the atmosphere (*as carbon dioxide*), plants, and animals. Study the following diagram to see how the **carbon cycle** works:

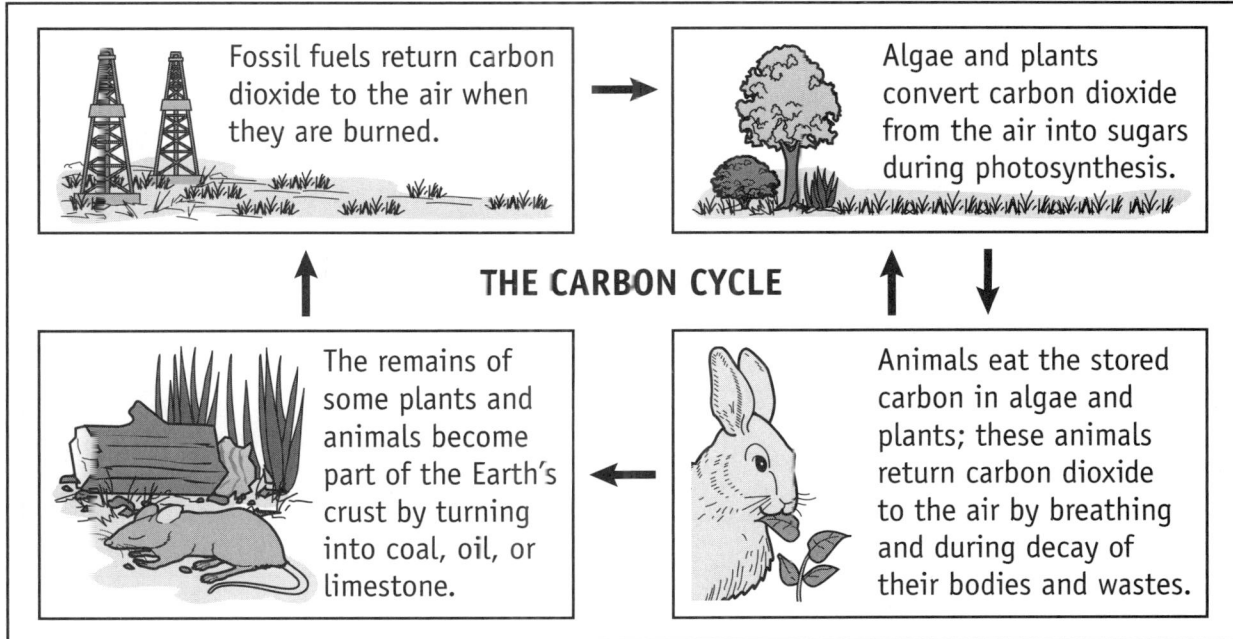

Fossil fuels return carbon dioxide to the air when they are burned.

Algae and plants convert carbon dioxide from the air into sugars during photosynthesis.

THE CARBON CYCLE

The remains of some plants and animals become part of the Earth's crust by turning into coal, oil, or limestone.

Animals eat the stored carbon in algae and plants; these animals return carbon dioxide to the air by breathing and during decay of their bodies and wastes.

THE NITROGEN CYCLE

Life on Earth also depends on the availability of nitrogen. Like the water and carbon cycles, the **nitrogen cycle** reuses the same material in various forms. Although nitrogen is plentiful in the air, it is not in a form that animals or plants can use. Instead, small organisms called bacteria turn this nitrogen into useful **nitrates** (*a nitrogen compound*).

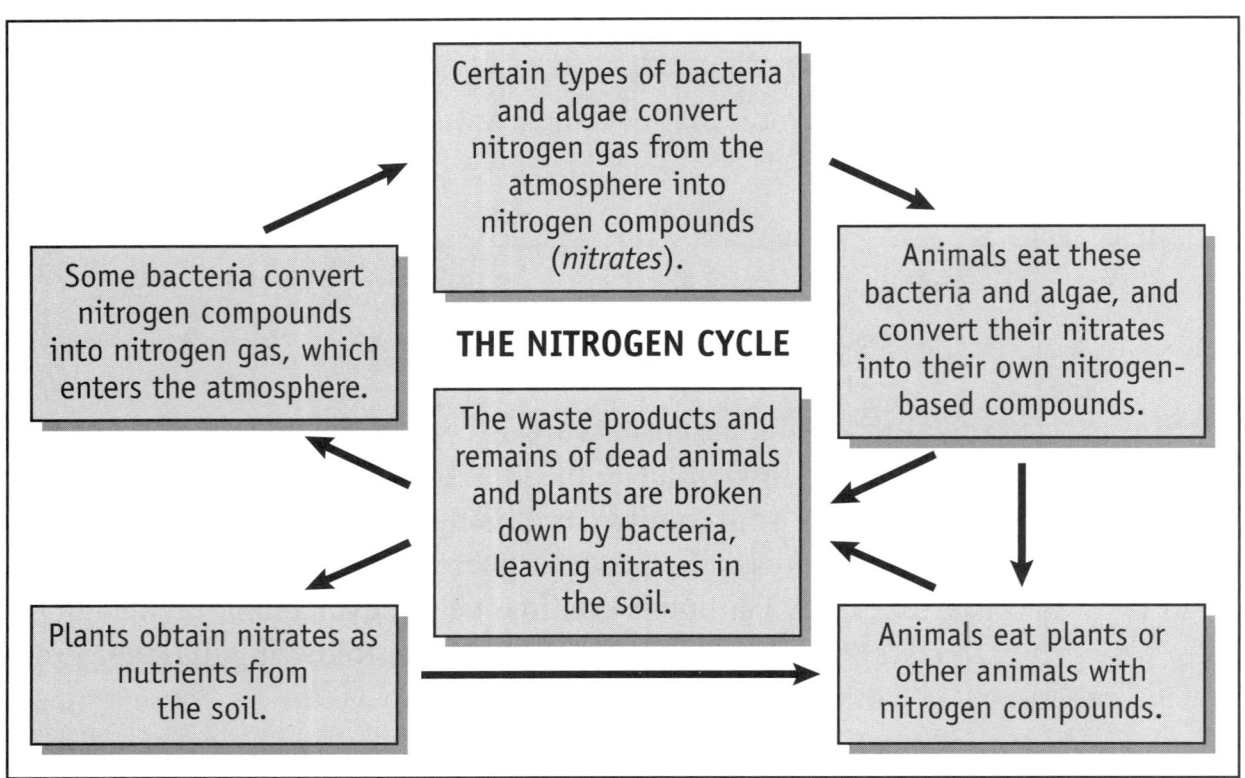

Some bacteria convert nitrogen compounds into nitrogen gas, which enters the atmosphere.

Certain types of bacteria and algae convert nitrogen gas from the atmosphere into nitrogen compounds (*nitrates*).

THE NITROGEN CYCLE

The waste products and remains of dead animals and plants are broken down by bacteria, leaving nitrates in the soil.

Animals eat these bacteria and algae, and convert their nitrates into their own nitrogen-based compounds.

Plants obtain nitrates as nutrients from the soil.

Animals eat plants or other animals with nitrogen compounds.

APPLYING WHAT YOU HAVE LEARNED

✦ "The carbon cycle depends on the air while the nitrogen cycle depends on the soil." Explain this statement. _____

_____

_____

_____

_____

Because different forms of life depend on the **water**, **carbon** and **nitrogen cycles**, problems can arise when human activities interfere with these cycles. For example, by burning fossil fuels such as coal, humans release more carbon dioxide into the air than plants can use. At the same time, humans cut down tropical rainforests, which reduces the number of plants conducting photosynthesis. When this happens, the carbon cycle is changed and greater amounts of carbon dioxide build up in the atmosphere.

Destruction of Amazon rain forest

APPLYING WHAT YOU HAVE LEARNED

Identify three results that might occur from modifying the Earth's water, carbon and nitrogen cycles.

Earth's Cycle	A Possible Result
Water Cycle	
Carbon Cycle	
Nitrogen Cycle	

COMPARING THE EARTH AND MOON

Now that you've learned about the Earth's materials, resources, and systems, you can appreciate how different our planet is from the moon. The moon is a rocky sphere without water or air. It is pockmarked by craters from meteor strikes and smooth areas from lava flows. There are no living things on the moon, and temperatures can change suddenly based on whether an area is in the sun or shade.

On Earth, the gases of our **atmosphere** act as a blanket holding in the heat and reducing extremes in temperature. Our atmosphere also acts as a cushion capturing evaporated water and sending it back to the Earth as rain and snow. The water, carbon and nitrogen cycles allow plants and animals to have the resources they need without using them up.

Most land forms on the Earth are covered with soil, which allows plants to send down roots. This soil allows plants to absorb water, nitrates, and other nutrients. Plants capture energy from the sun through photosynthesis. Animals eat these plants to survive. Finally, the dead bodies of plants and animals decay and enrich the soil, where other plants absorb their nutrients.

APPLYING WHAT YOU HAVE LEARNED

✦   Imagine you are an astronaut sent to the moon to establish the first human colony there. What steps would need to be taken to make life there possible?

_____

_____

_____

_____

WHAT YOU SHOULD KNOW

A. You should know that the Earth has **renewable**, **nonrenewable**, and **inexhaustible resources**.

B. You should know that past events have led to the formation of many of the Earth's resources.

C. You should know that the Earth's different systems interact with each other. The Earth's oceans, atmosphere and land forms all influence each other. Weather patterns are one example of this interaction.

D. You should know that Earth's water, carbon and nitrogen supplies are continually recycled in its **water**, **carbon** and **nitrogen cycles**.

E. You should know that because of its atmosphere, soil, oceans, and living things, Earth's physical characteristics are different from those of the moon.

CHAPTER STUDY CARDS

Types of Natural Resources

★ **Renewable Resources.** Resources that can be replaced in a short time, such as forests, livestock, and food crops.

★ **Nonrenewable Resources.** Resources formed over long time periods and can't be replaced, such as oil, coal, and minerals.

★ **Inexhaustible Resources.** Resources available in large supply and can't be used up, such as solar energy and wind energy.

Interactions of Earth's Systems

Earth's land forms, oceans, and atmosphere interact with each other.

★ **Ocean and Land Forms.** Ocean currents can erode coastlines.

★ **Weather Patterns.** Weather is the condition of the atmosphere; climate is typical weather over a long period.
  • Tornadoes: High-speed wind patterns.
  • Hurricanes: Heavy rains and strong winds; start in ocean in tropical areas.

Earth's Cycles

★ **Water Cycle.** Earth's water is constantly moving in and out of the atmosphere. Process begins with sun's heat causing water to evaporate into air. Water returns as precipitation.

★ **Carbon Cycle.** Carbon is continually being recycled between atmosphere, (carbon dioxide), plants and animals.

★ **Nitrogen Cycle.** Life also depends on nitrogen, which plants get from soil.

Comparison of the Earth and Moon

★ **Earth.** Its atmosphere acts like a blanket holding in heat and preventing extremes in temperature. Has a water, carbon and nitrogen cycle. Land forms are covered with soil. Has plant and animal life.

★ **Moon.** Surface is rocky and crater-filled. Lacks water and air. Has no plant or animal life. Temperatures change suddenly from day to night.

CHECKING YOUR UNDERSTANDING

1 Which part of the water cycle came before the X in the diagram?

OBJ. 4
5.6 (B)

HINT   To answer this question, you need to understand the water cycle: water evaporates into the atmosphere where it forms droplets. These droplets make clouds. Before the water was in the cloud in the atmosphere, it was evaporating from the Earth's surface. Thus, the best answer is **D**.

Now try answering some additional questions on your own:

2 A student looks outside her window after it has rained and sees puddles of water. Later that same day, the puddles are gone. What process explains why the puddles have disappeared?

F Runoff
G Precipitation

H Evaporation
J Condensation

OBJ. 4
5.6 (B)

3 Which of the following is an example of condensation in the water cycle?

A Clouds forming in the atmosphere
B Water drops falling through the air
C Streams flowing into rivers
D Puddles disappearing on a hot day

OBJ. 4
5.6 (B)

4 In which part of the diagram is water changing from a liquid to a gas?

F A
G B
H C
J D

OBJ. 4
5.6 (B)

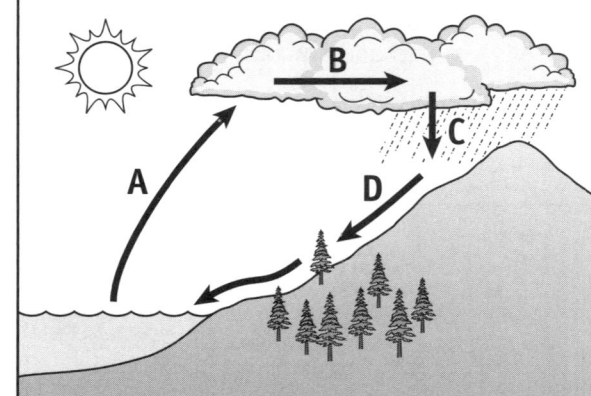

5 Trees, wild flowers, and grasses are all considered to be —

A renewable resources
B nonrenewable resources

C inexhaustible resources
D artificial resources

OBJ. 4
3.11 (A)

6 Which of the following would be considered a renewable resource?

F Wood, because new trees can grow where others were cut down
G Coal, because more can be made in 100 million years
H Petroleum, because it can be refined into gasoline
J Gold, because it can be mined from the Earth

OBJ. 4
3.11 (A)

7 **What step is missing from the diagram on the right?**

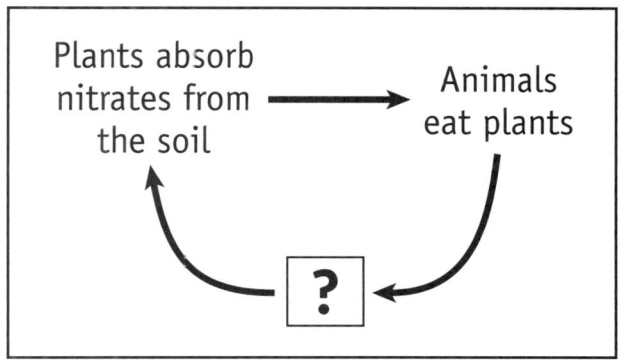

A Animals release nitrogen gas.

B Plants absorb nitrogen from the air.

C Dead plants and animals return nitrates to the soil.

D Rainfall produces more nitrates.

OBJ. 4
5.6 (B)

8 **During photosynthesis, plants make sugars, which contain carbon from the atmosphere. Animals eat plants and absorb these sugars. Later, animals burn these sugars for energy. What gas do animals produce that puts carbon back into the atmosphere?**

F Oxygen

G Carbon dioxide

H Nitrogen

J Water vapor

OBJ. 4
5.7 (A)

The diagram below shows a place where air currents form due to the uneven heating of the Earth.

9 **In which direction will air currents most likely move?**

OBJ. 4
4.11 (B)

A Air over the sea will move away from the land.

B Air will move from the land toward the sea.

C Air over the land will rise.

D Air over the sea will rise.

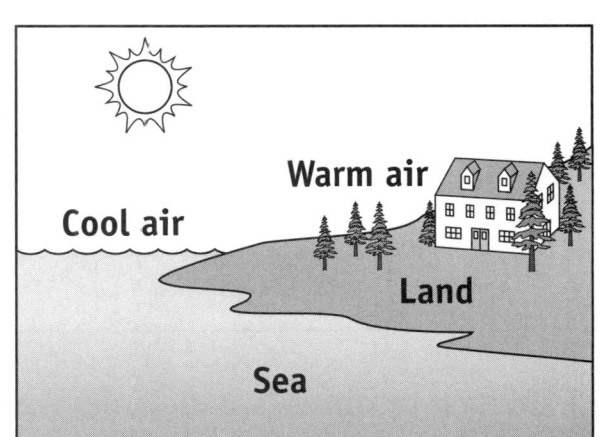

10 **Certain storms form over water near the equator, usually between June and November. Warm moist air rises over the ocean, causing a strong, swirling storm with high winds and heavy rains. What type of storm is this describing?**

F Tornado

G Hurricane

H Snowstorm

J Sandstorm

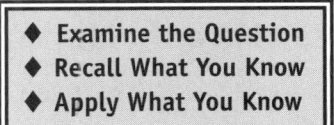

◆ Examine the Question
◆ Recall What You Know
◆ Apply What You Know

OBJ. 4
4.6 (A)

11 **What type of weather is likely to cause the most erosion over a long period of time?**

A Warm, sunny weather

B Cool, dry temperatures

C Little or no wind

D Heavy, constant rainfall

OBJ. 4
5.11(A)

The diagram below shows a water cycle. Four parts of the diagram have been labeled A, B, C, and D.

12 **Which best describes what is happening in the water cycle at B?**

F Water from the land is returning to the oceans. *OBJ. 4 5.6 (B)*

G Ocean water is evaporating into the atmosphere.

H Water from the atmosphere is returning to Earth's surface.

J Water vapor in the atmosphere is condensing into clouds.

WATER CYCLE

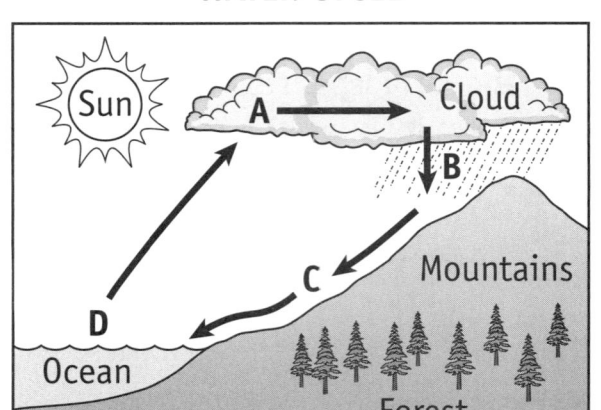

Renewable Resources	Nonrenewable Resources	Inexhaustible Resources
Trees	Oil	Fresh water
Soil	Coal	Sunlight

13 **Which resource is under the wrong heading?**

A Trees

B Fresh water

C Soil

D Oil

◆ Examine the Question
◆ Recall What You Know
◆ Apply What You Know

OBJ. 4
3.11(A)

14 **The oil that is burned today was formed from —**

F minerals in the Earth

G the remains of organisms that lived millions of years ago

H chemical reactions that took place recently

J volcanic ash that mixed with ocean waters

OBJ. 4
5.11(C)

15 **Which of the following is considered a fossil fuel?**

A Oil   C Water

B Gold   D Nitrogen

OBJ. 4
5.11(C)

16 **The diagram shows a mountain range along an ocean coastline. Location X will probably have —**

F warmer summers than location Y

OBJ. 4
4.11(B)

G more rainfall than location Y

H less wind than location Y

J colder winters than location Y

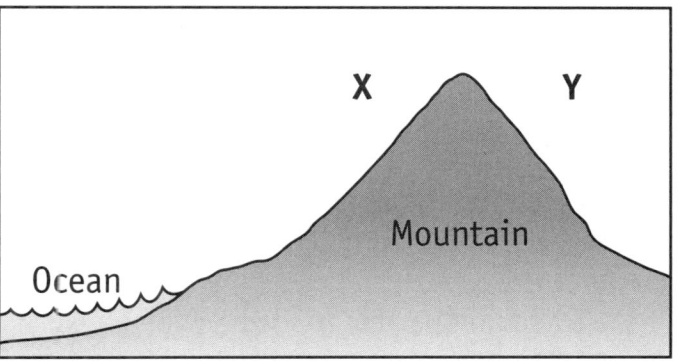

17 **Compared to a nearby inland area, a land area next to the ocean will most likely have —**

A cooler summers   C warmer summers

B drier summers   D less windy summers

OBJ. 4
4.11(B)

18 **Which of these resources for home building is renewable?**

F Oil   H Lumber

G Aluminum   J Copper

OBJ. 4
3.11(A)

19 **The energy that causes water on the surface of the ocean to turn to water vapor comes from the —**

A motion of animals   C clouds

B underwater volcanoes   D sun

OBJ. 4
4.11(C)

20 **Which physical characteristic is found on Earth but *not* on the moon?**

F Heat energy from the sun

G Large and rocky areas

H Rich soil from decayed plant life

J Temperature changes between day and night

◆ Examine the Question
◆ Recall What You Know
◆ Apply What You Know

OBJ. 4
5.12(C)

21 **How many times in a day do tides usually occur on the Texas coast?**

A None   C Twice

B Once   D Four

OBJ. 4
5.6(A)

CHECKLIST OF OBJECTIVES IN THIS UNIT

Directions. Now that you have completed this unit, place a check (✔) next to those objectives you understand. If you are having trouble recalling information about any of these objectives, review the chapter listed in the accompanying brackets.

- [ ] You should be able to interpret how land forms result from a combination of constructive and destructive forces such as the build-up of sediment and weathering. **[Chapter 10]**
- [ ] You should be able to identify the physical characteristics of Earth and compare them to the physical characteristics of the moon. **[Chapters 9 and 11]**
- [ ] You should be able to identify and observe actions that require time for changes to be measurable, including growth, erosion, dissolving, weathering, and flow. **[Chapter 10]**
- [ ] You should be able to draw conclusions about "what happened before" using data such as from tree-growth rings and layers of sedimentary rock. **[Chapter 10]**
- [ ] You should be able to identify past events that led to the formation of the Earth's renewable, nonrenewable, and inexhaustible resources. **[Chapter 11]**
- [ ] You should be able to identify events and describe changes that occur on a regular basis such as in daily, weekly, lunar, and seasonal cycles. **[Chapters 9 and 11]**
- [ ] You should be able to identify the significance of the water, carbon, and nitrogen cycles. **[Chapter 11]**
- [ ] You should be able to test properties of soils including texture, capacity to retain water, and ability to support life. **[Chapter 10]**
- [ ] You should be able to summarize the effects of the oceans on land. **[Chapter 11]**
- [ ] You should be able to identify the sun as the major source of energy for the Earth and understand its role in the growth of plants, in the creation of winds, and in the water cycle. **[Chapter 11]**
- [ ] You should be able to identify and describe the importance of Earth's materials, including rocks, soil, water, and gases of the atmosphere and be able to classify them as renewable, nonrenewable, or inexhaustible resources. **[Chapters 10 and 11]**
- [ ] You should be able to identify the planets in our solar system and their position in relation to the sun. **[Chapter 9]**
- [ ] You should be able to describe the characteristics of the sun. **[Chapter 9]**
- [ ] You should be able to identify how the surface of the Earth can be changed by forces such as earthquakes and glaciers. **[Chapter 10]**
- [ ] You should be able to describe some cycles, structures, and processes that are found in a simple system and describe some of their interactions. **[Chapters 9, 10 and 11]**
- [ ] You should be able to identify patterns of change such as in weather and objects in the sky. **[Chapters 9 and 11]**

UNIT 5

A PRACTICE ELEMENTARY SCIENCE TAKS

This chapter consists of a complete practice **Elementary Science TAKS**. Before you begin, let's review a few directions for the test:

★ **Answer All Questions.** The TAKS in Science consists of 50 multiple-choice questions. Do not leave any questions unanswered since there is no penalty for guessing. Blank answers are counted as wrong.

★ **Use the "E-R-A" Approach.** Remember to carefully **examine** the question to understand what the question is asking. Next, **recall** what you have learned about that particular topic in science. Finally, **apply** your knowledge to answer the question.

★ **Use the Process of Elimination.** When answering a multiple-choice question, it may be clear that certain choices are wrong. After you eliminate incorrect choices, select the *best* response that remains.

★ **Revisit Difficult Questions.** If you run into a difficult question, do not be discouraged. Put a check mark (✔) next to it. Answer it as best you can and move on to the next question. At the end of the test, go back and reread any questions you marked. Sometimes the answer to a difficult question might become clearer to you later.

★ **Read Items Carefully.** Read the directions carefully. Be sure to examine all parts of the question. If you have time left at the end, check your work and correct any errors.

★ **When You Finish.** When you finish the test, make sure you have answered all the questions. Also make sure that your answers correspond to the correct question number if you use an answer sheet. Do not disturb other students!

This practice test lists the TAKS Objective and grade level number tested by each question. This will help you identify any topics you may still need to study.

Good luck on this practice test!

Centimeters

CHAPTER 12

A PRACTICE TAKS IN SCIENCE

Directions: Read each question carefully. Then circle or write down the letter of the choice that is the correct answer.

1 **The illustration on the right shows a tropical rainforest. What gas in the atmosphere would increase if the trees in this rainforest were cut down?**

A   Carbon dioxide
B   Water vapor
C   Oxygen   OBJ. 4
             5.6 (B)
D   Nitrogen

2 **What happens to the dead leaves and animals that fall to the forest floor?**

F   They decay and become part of the soil.
G   They disintegrate into the atmosphere.   OBJ. 4
                                                 5.11 (C)
H   They remain preserved in the rainforest.
J   They are washed into the oceans by heavy rains.

3 **How do the trees in this rain forest obtain their energy?**

A   By eating other plant life
B   By using energy from the sun   OBJ. 2
                                         2.9 (B)
C   By drinking water through their roots
D   By absorbing heat from the forest floor

Mineral	Color	Streak
Anhydrite	Colorless, white, gray, blue or violet	White
Quartz	Colorless, white, purple, or gray	White
Graphite	Black to silver	Black to brownish gray
Hematite	Silver gray, black, red or brown	Red or brown

4   **Using the table above, what is the most likely streak color of halite, a mineral that is colorless, white, blue or purple?**

OBJ. 1
5.2 (D)

   F   Brown

   G   Black

   H   White

   J   Red

5   **Which scientific conclusion is best supported by evidence from the graph?**

OBJ. 1
5.3 (A)

   A   Gravity varies with the mass of the planet.

   B   Gravity increases the closer a planet is to the sun.

   C   Gravity is the same throughout our solar system.

   D   Gravity moves the planets away from the sun.

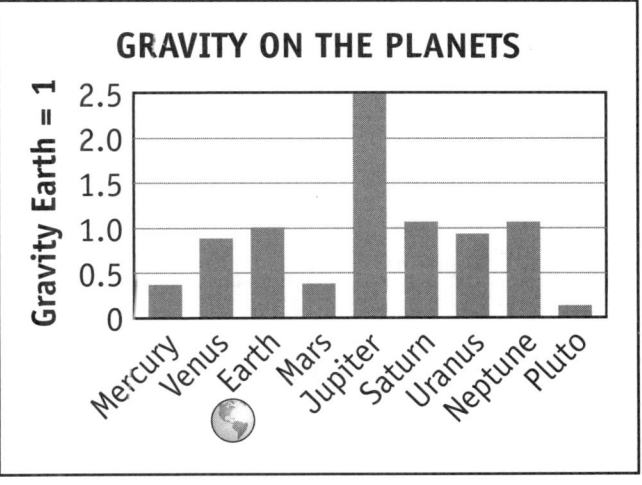

6   **The Australian echidna, or spiny anteater, lives in the desert and eats ants. Which characteristic would NOT be an advantage to the echidna in such an environment?**

   F   Long, sticky tongue

OBJ. 2
2.9 (A)

   G   Long and narrow snout

   H   Keen sense of smell

   J   Large teeth

The Spiny Anteater

7 Students observe cars stopping on a rough gravel road and on a smooth, paved roadway. What conclusion are they most likely to draw from their observations?

    A Both roads have the same friction.

    B The paved road has greater friction.

    C The gravel road has greater friction.

    D At different times, each road has greater friction.

OBJ. 3
3.6 (A)

8 Why do manufacturers often put plastic handles on metal pots?

    F Both plastic and metal conduct heat well.

    G Plastic does not conduct heat as well as metal does.

    H Metal does not conduct heat as well as plastic does.

    J Neither metal nor plastic conduct heat well.

OBJ. 3
5.7 (A)

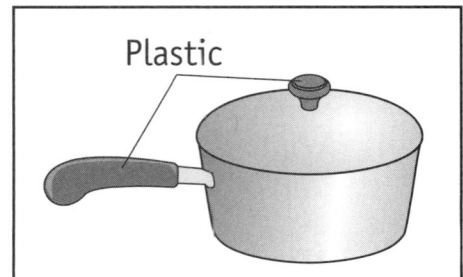

Plastic

9 While playing basketball, a student throws a ball into the air. What force brings the basketball back to the ground?

    A Gravity

    B Magnetism

    C Friction

    D Conductivity

OBJ. 3
3.6 (A)

10 Which of the following changes electrical energy into light energy?

OBJ. 3
5.8 (C)

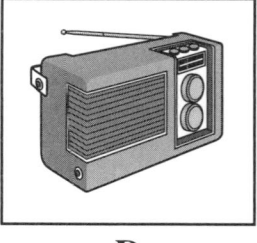

    A   **B**   **C**   **D**

11 Light bouncing off a mirror is —

    F refracted

    G transmitted

    H absorbed

    J reflected

OBJ. 3
5.8 (B)

12 **The sun can best be described as —**

F   an icy planet

G   a ball of superheated gases

H   a cloud of dust

J   a burned-out star

OBJ. 4
3.11 (D)

13 **What happens if Bulb A burns out in this circuit?**

OBJ. 3
5.8 (C)

A   Bulb B will remain lit.

B   Bulb B will also go out.

C   The wires will become very hot.

D   The battery will lose its electricity.

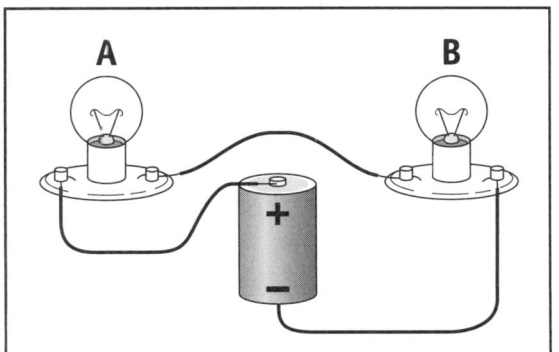

14 **Which of these materials is a good conductor of electricity?**

F   Rubber

G   Glass

H   Copper

J   Newspaper

OBJ. 3
5.7 (A)

15 **Tiny drops of water are light enough to float in the air as clouds. As these drops of water bump into each other, they combine into larger drops of water. What happens when these drops become too large to float in the air?**

A   They collect underground.

B   They fall as rain.

C   They evaporate.

D   They create fog.

OBJ. 4
5.6 (B)

16 **Bottlenose dolphins live in the water. They look like fish but breathe air like people. They send out special sounds and wait for the echo to tell where things are. Which sense is most useful for dolphins in finding objects underwater?**

F   Taste

G   Smell

H   Hearing

J   Touch

OBJ. 2
5.9 (A)

A bottlenose dolphin

17 **Which two planets are closest to Saturn?**

A Jupiter and Uranus
B Venus and Earth
C Mars and Mercury
D Mercury and Neptune

OBJ. 4
3.11 (C)

Saturn

A class recorded the outdoor temperatures at noon on the first day of each month for the entire school year. They created the following chart with their information:

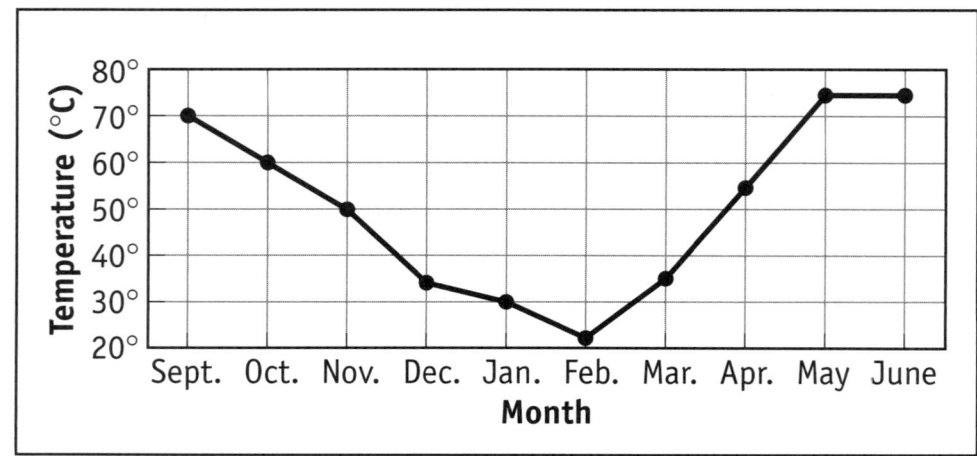

18 **What was the temperature on the first day of April?**

F 0°   H 55°
G 30°   J 75°

OBJ. 1
5.2 (C)

19 **From November to February, the outdoor temperature —**

A decreased   C remained the same
B increased, then decreased   D increased

OBJ. 1
5.2 (C)

 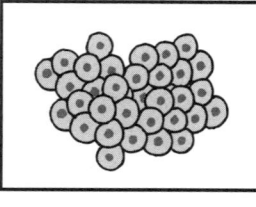  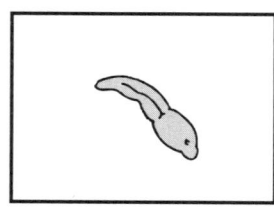

1   2   3   4

20 **Which is the correct order of development for this frog?**

F 4, 2, 3, 1   H 3, 1, 4, 2
G 2, 1, 4, 3   J 2, 4, 1, 3

OBJ. 2
5.6 (C)

21 **Field mice learn some behaviors while interacting with the environment. Which behavior is most likely learned?**

A Drinking water
B Running fast
C Avoiding humans
D Sleeping during the day

OBJ. 2
5.10 (B)

22 **Over time, a forest region gradually changes so that it has colder winters with heavy snowfalls. Which group of rabbits would most likely benefit from this change to the environment?**

F Rabbits with dark fur
G Rabbits with sharp claws
H Rabbits without claws
J Rabbits with white fur

OBJ. 2
3.8 (C)

A science class was learning about daily temperatures. The following chart shows the temperatures recorded by the students in the class during one week in June.

TEMPERATURES FOR THE WEEK

Day of the Week	Temperature in Degrees
Monday	72° F
Tuesday	76° F
Wednesday	68° F
Thursday	70° F
Friday	70° F

OBJ. 1
5.4 (A)

23 **Which thermometer shows the temperature recorded on Friday?**

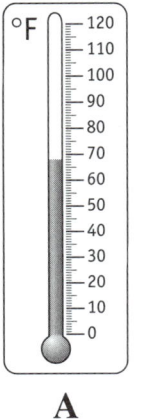

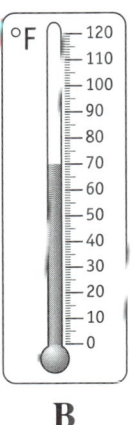

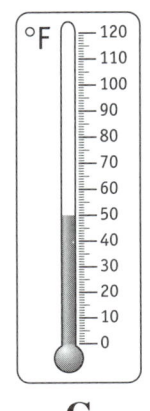

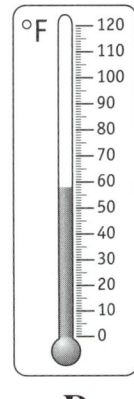

A   **B**   **C**   **D**

24 If the device illustrated to the right were used in a laboratory, what would it primarily be used for?

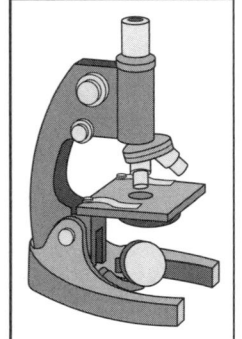

F   To measure changes in temperature
G   To read the amount of rainfall
H   To see objects from a distance
J   To see larger views of small objects

OBJ. 1
5.4 (A)

25 What explains Earth's changing seasons?

A   The moon's orbit around Earth
B   Earth's tilt as it revolves around the sun
C   Earth's spinning around its axis
D   The effect of nearby planets on Earth

OBJ. 4
5.6 (A)

A student conducted an experiment to find out how temperature affects air in a balloon. He drew a line around the center of the balloon and measured the length of the line around the balloon. The results are recorded in the chart:

26 What conclusion could be drawn from the information in this chart?

F   The warmer the balloon gets, the larger it becomes.
G   The balloon is unaffected by changes in temperature.
H   The balloon is larger in the freezer than outdoors.
J   The warmer the balloon gets, the smaller it becomes.

OBJ. 1
5.2 (C)

HOW TEMPERATURE AFFECTS AIR IN A BALLOON

Conditions of Balloon	Length of Line Around Balloon (in centimeters)
Balloon after coming out of the freezer	14 cm
Balloon at room temperature	21 cm
Balloon after being warmed for 2 min	33 cm
Balloon after being warmed for 4 min	54 cm

27 Which action can be taken to prevent soil erosion?

A   Planting grass
B   Raising grazing animals
C   Digging ditches
D   Cutting down trees

OBJ. 4
5.12 (A)

28 A fifth grade class recorded the high and low temperatures in their class each day for a week. They created a table of their results. On which day of the week did the temperature change the most?

F Monday
G Wednesday
H Thursday
J Friday

OBJ. 1
5.2 (C)

TEMPERATURE READINGS AT WITNEY ELEMENTARY SCHOOL

	Low Temperature (°C)	High Temperature (°C)
Monday	19	28
Tuesday	20	27
Wednesday	20	28
Thursday	18	26
Friday	19	29

29 Which of these materials would be attracted to a magnet?

A A gold ring
B A stainless steel fork
C A copper penny
D A glass lens

OBJ. 3
5.7 (A)

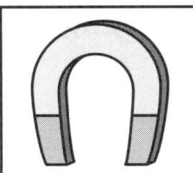

Use the following diagram to answer questions 30 and 31

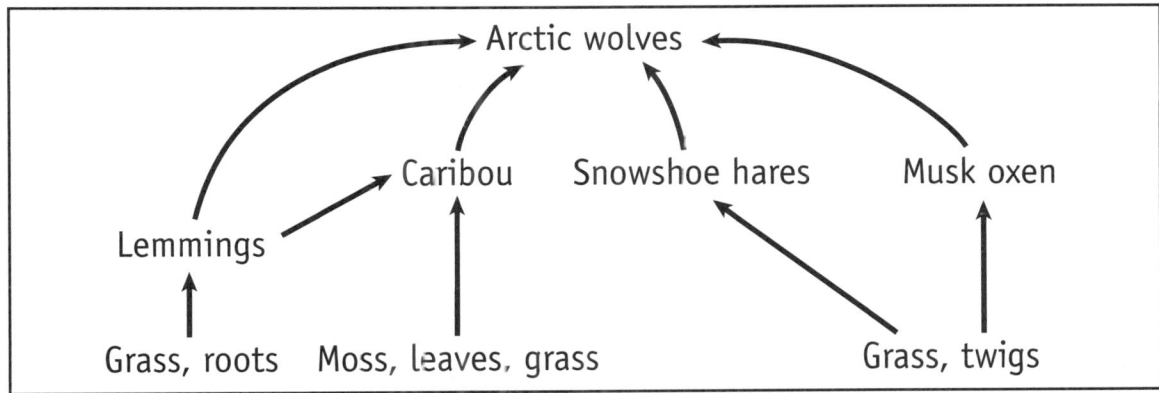

30 According to the food web shown above, an increase in the number of arctic wolves would probably result in —

F a decrease in the amount of grass
G an increase in the number of caribou
H an increase in the number of snowshoe hares
J a decrease in the number of musk oxen

OBJ. 2
2.9 (B)

31 Which animal in this food web is shown as both a predator and prey?

A Caribou
B Lemming
C Snowshoe hare
D Arctic wolf

OBJ. 2
5.9 (B)

Renewable Resources	Inexhaustible Resources
Fresh water	Sunlight
Trees	?

32 **Which of these best completes the above chart?**

OBJ. 4
3.11(A)

    F   Oxygen gas
    G Wind energy
    H Fertile soil
    J   Coal

33 **Using a centimeter ruler, measure the width of this carrot from point A to point B to the nearest centimeter. Record and bubble your answer in the grid.**

OBJ. 1
5.2(B)

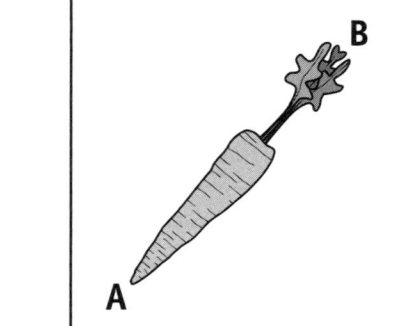
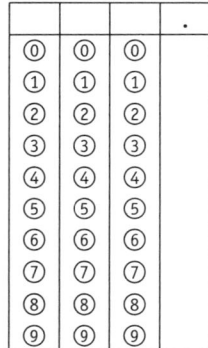

34 **A student dissolves 10 grams of salt in a beaker with 1 liter of water. How can the student get the salt back?**

OBJ. 3
5.7 (B)

    F   Pour the solution through a paper filter
    G Place a magnet over the beaker
    H Shine a bright light through the beaker
    J   Boil the solution until the water evaporates

35 **The process illustrated below shows water in a container. What process is taking place?**

    A Melting
    B Condensation
    C Freezing
    D Evaporation

OBJ. 3
5.7 (B)

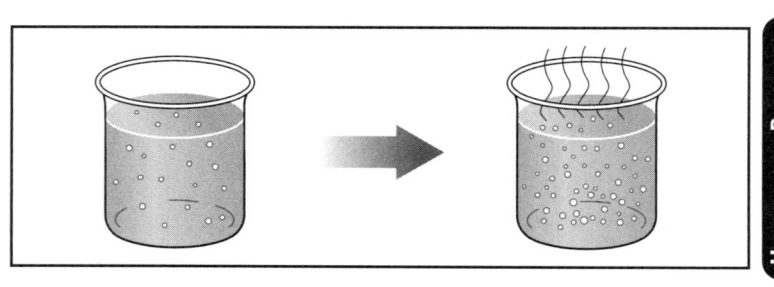

Object	Paper plane	Baseball	Drinking glass	Iron pan
Mass	10g	50g	80g	500g

36 **Which object requires the most force to move a distance of 10 meters?**

F   Paper plane
G   Baseball
H   Drinking glass
J   Iron pan

OBJ. 3
3.6 (A)

37 **In a model of the solar system, a light bulb is used to represent the sun. A tennis ball represents the moon, and a marble is used to stand for the Earth. How could the model be made more accuate?**

A   Use a light bulb to represent the moon.
B   Use a tennis ball to represent the moon and sun.
C   Use the marble to represent the moon and the tennis ball for the Earth.
D   Use the light to represent the Earth and the tennis ball to represent the sun.

OBJ. 1
5.3 (C)

38 **A graduated cylinder contains 30 mL of water. A pebble is then placed in the cylinder. The diagram shows the volume of water after the pebble is placed in the cyclinder. What is the volume of the pebble?**

F 5 cm^3
G 15 cm^3
H 40 cm^3
J 55 cm^3

OBJ. 1
5.2 (B)

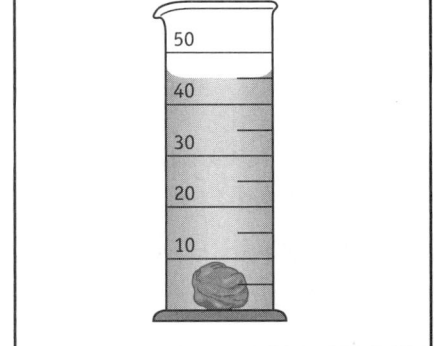

39 **A student places a plastic button on a table 1 cm from a strong magnet. The student observes if the button moves. The student removes the button and places a paper clip 1 cm from magnet. The student observes whether the paper clip moves. What question is the student most likely trying to answer?**

A   Is magnetism caused by gravity?
B   Which materials are magnetic?
C   Can a magnet work underwater?
D   Does light affect the power of a magnet?

OBJ. 3
5.7 (A)

40 **What do the eyespots on this butterfly's wings probably help it to do?**

OBJ. 2
2.9 (A)

   F Stay warm
   G Frighten off enemies
   H Fly away quickly
   J Locate nectar-filled flowers

41 **Two students wanted to find out which of their toy trucks would go farthest. They decided to let each truck roll down a ramp and then measure how far it rolled on the ground.**

Which of these should be held constant if they want a fair test?

   A The height of the ramp
   B The time of day
   C The temperature of the room
   D The weight of the ramp

OBJ. 1
5.2 (A)

42 **An electric company wants to make space for a new powerline. Trees and bushes are cleared from a strip of land it owns. What impact is this action likely to have on birds living in the area?**

   F Less food and shelter will be available.
   G Weather conditions will change.
   H The rate of erosion will greatly decrease.
   J The natural enemies of the birds will increase.

OBJ. 2
3.8 (C)

43 **Which is an important safety rule in a laboratory experiment in which chemicals will be mixed?**

   A Keep all windows closed
   B Wear safety goggles
   C Turn out all electric lights
   D Use a fire blanket

OBJ. 1
5.1 (A)

44 **Which of the following characteristics of Earth is also found on the moon?**

   F Soil
   G Atmosphere
   H Water
   J Gravity

OBJ. 4
5.12 (C)

Use the following illustrations to answer question 45

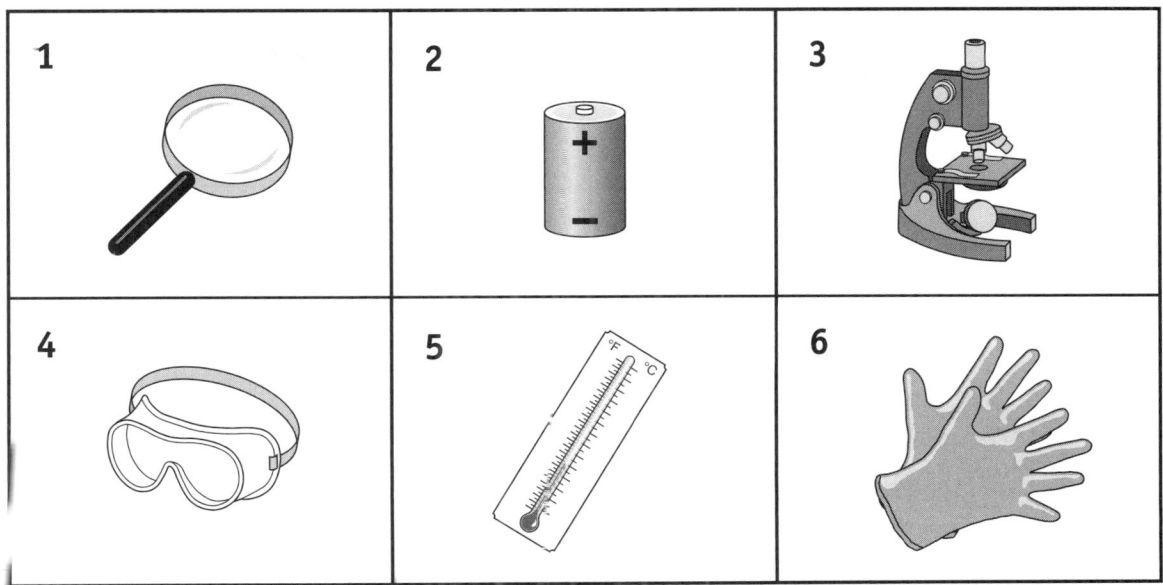

45 **Which pieces of laboratory equipment are used for safety?**

A   1 and 2
B   3 and 5
C   3 and 4
D   4 and 6

OBJ. 1
5.1 (A)

46 **Which is an inherited trait of a chameleon?**

F   Missing a leg from an accident
G   Preferring to drink water from streams
H   An ability to change its skin color
J   Living in a particular rainforest

OBJ. 2
5.10 (A)

47 **Which of the following shows a process that builds up Earth's landforms?**

OBJ. 4
5.12 (A)

A   B   C   D

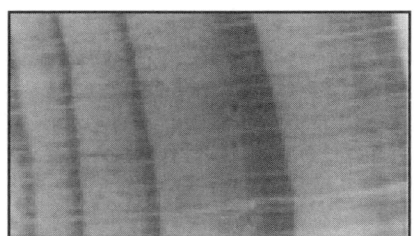
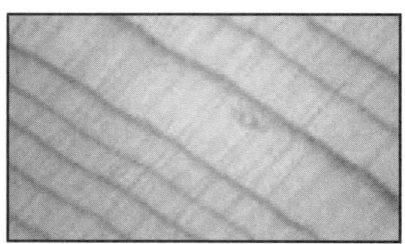

48 These photographs show tree-growth rings from two different trees grown in the same area. What conclusion can be drawn from these tree-growth rings?

   F Rainfall was generally even in this area.

   G Some past years had more rainfall than others.

   H These trees grew in an environment without rainfall.

   J These trees were killed by flooding.

OBJ. 4
5.11 (B)

49 Most of the mass of one serving of the cereal Miller's Oats is —

   A fat

   B sugar

   C carbohydrate

   D protein

OBJ. 1
5.3 (B)

MILLER'S OATS Nutrition Facts		
Serving Size 1/2 cup (40g)		
Amount Per Serving		% Daily Value
Total Fat 3g		4%
Cholesterol 0g		0%
Total Carbohydrates 27g		9%
Dietary Fiber 4g		5%
Sugar 1g		--
Protein 5g		--

Sandy soil

Soil with high clay content

Soil with high content of decayed plant material

1   2   3

50 Students collected the three samples of different types of soils shown above. They inspected the three samples with a hand lens and rubbed each sample with their hands. What question was this investigation designed to answer?

   F How well does each type of soil support life?

   G How well does each type of soil hold water?

   H What is the texture of each type of soil like?

   J What microscopic bacteria live in the soil?

OBJ. 4
4.11(A)

A

Adaptive Characteristic. The quality of a plant or animal that helps it survive in its environment. [67]

Animal. A living organism that can move freely, but cannot produce its own food. It eats plants and other animals to survive. [57]

Atmosphere. The mixture of gases surrounding the Earth. The atmosphere protects the Earth from ultraviolet rays, cushions the Earth from meteors, holds water, spreads out the sun's heat, and provides oxygen to animals and carbon dioxide to plants. [169]

B–C

Balance. A dual-pan or triple-beam balance used to measure mass. [25]

Boiling Point. The temperature at which a liquid will turn into a gas. [93]

Carbon Cycle. A cycle in which carbon dioxide is produced by animals or the burning of fossil fuels, goes into the air, is used by plants to make carbon compounds during photosynthesis, and then is absorbed by animals when they eat the plants. [167]

Carbon Dioxide. A gas in the atmosphere. Carbon dioxide is produced by animals and used by plants for photosynthesis. [168]

Carnivore. An animal, such as a wolf, that eats other animals. [72]

Centimeter. A distance measuring one-hundredth of a meter. [22]

Collecting Net. A net used to collect samples in field investigations [13]

Conductivity. The ability of matter to carry heat, sound, or electricity. [95]

Constructive Forces. Forces that build up Earth's surface, such as sedimentation, lava flows or the folding of the Earth's crust into mountains. [145]

Cycle. A series of steps in which the last step leads back to the first step, such as the water cycle or the carbon cycle. [165]

D–E

Day / Night. Cycle of light and darkness occurring every 24 hours, caused by the Earth's rotation on its axis. [133]

Destructive Forces. Forces that tear down Earth's land forms, such as weathering, erosion, glaciers, and earthquakes. [146]

Earthquake. A movement of Earth's crust. Can act to break down or build up land forms. [149]

Ecosystem. All the living and non-living things in an area; they depend on each other. [65]

Electrical Circuit. A closed path that electricity follows. Any break in the path will stop the flow of electricity throughout the entire path. [116]

Energy. Something with the power to do work. Energy takes different forms. Electricity, light, and heat are all different forms of energy. [112]

Erosion. The process by which rock and soil are worn down and carried away by wind, running water, ocean waves or glaciers. [147]

F–G

Food Chain. A chain showing the flow of energy between different plants and animals; a food chain shows what eats what. [73]

Food Web. A diagram showing the interaction of several food chains. [74]

Force. A push or pull acting on an object that will cause that object to move or change movement. [109]

Fossil. An impression left in sedimentary rock by the remains of a dead plant or animal. [151]

Fossil Fuels. Materials made from the remains of living organisms millions of years ago that can now be burned for energy. Oil, coal, and natural gas are fossil fuels. [162]

Friction. The force caused by the rubbing together of two or more surfaces, which slows down the motion of objects. [110]

Gas. A state of matter without any fixed shape or volume, such as oxygen gas. [93]

Glacier. A giant sheet of slow-moving ice. As glaciers move, they scrape the Earth's surface, digging holes and smoothing mountains. [148]

Graduated Cylinder. A cylinder used to measure the volume of liquids. [12]

Gravity. The force pulling objects to Earth. Gravity is a form of attraction between any two objects. The greater the mass of a planet is, the stronger its force of gravity will be. [111]

H–I

Habitat. The area in an ecosystem that a particular type of organism makes its home, such as tree tops in a rainforest canopy. [66]

Hand Lens. A lens that magnifies objects by several times, but not as powerful as a microscope. [12]

Herbivore. An animal, such as a cow, that eats only plants. [72]

Inexhaustible Resource. Resources that cannot be used up, like sunlight or wind energy. [159]

Inherited Trait. A physical characteristic or natural instinct that a plant or animal inherits from its parents. An inherited trait cannot be changed. [82]

Insulator. A material that does not conduct something, such as electricity, well. Instead, it stops the flow of energy. For example, plastic is a good insulator. [96]

J–L

Kilogram. A mass of 1,000 grams. [22]

Kilometer. A distance of 1,000 meters. [21]

Lava Flow. Flow of molten rock coming from below Earth's surface. [146]

Learned Behavior. A behavior that an animal learns by interacting with its environment. [83]

Life Cycle. Changes that a plant or animal goes through in its life, from its beginning to its death. For example, a plant begins as a seed. If the seed germinates, it becomes a seedling; then it grows into a plant, eventually it dies. [58]

Light. A form of energy that can travel through empty space. [113]

Liquid. A state of matter where particles move more quickly than a solid; a liquid, such as water, no longer has a fixed shape but still has a fixed volume. [92]

Liter. A measure of the volume of a liquid. [22]

M

Magnet. A piece of iron or other material that can pull certain metals towards it, or pick them up. [94]

Mass. The amount of matter something has; measured in grams or kilograms. [90]

Matter. Anything that has mass and takes up space. [90]

Melting Point. The temperature at which a solid turns into a liquid. [93]

Metamorphosis. A process in which an animal completely changes its form during its life cycle. For example, a tiny egg becomes a tadpole and then a frog. A butterfly becomes a larva, then a pupa, and finally an adult butterfly. [58]

Microscope. A laboratory instrument used to magnify extremely small objects so that we can see them. [12]

Milliliter. A measure of the volume of a liquid, equal to one-thousandth of a liter. [22]

Millimeter. A distance measuring one-thousandth of a meter. [21]

Mixture. When two things, like sand and salt, are mixed together. The ingredients in a mixture often keep many of their properties, and can be separated by filters, the boiling of liquids, or similar physical methods. [97]

Model. Something that is made to represent something else, such as different balls representing the Solar System. Models can always be improved by making the size of the objects and distances more accurate. [40]

Motion. When an object changes its position over time. [107]

N–O

Niche. The unique place that each plant and animal has in an ecosystem, based on its unique characteristics and the role it plays in that ecosystem. [75]

Nitrogen Cycle. A cycle in which plants take nitrogen compounds from the soil, animals eat the plants, and nitrogen compounds are returned to the soil when the plants and animals die. [168]

Nonrenewable Resource. Resources formed over a very long period of time that cannot be regrown or replaced, such as a fossil fuel (oil or coal). [159]

Organism. Any living thing. [65]

Oxygen. One of the gases in the atmosphere. Animals need oxygen to live. Plants produce oxygen during photosynthesis. [57]

P–Q

Photosynthesis. A process by which plants capture energy from sunlight and convert this energy into food. [54]

Planets. Large bodies of rocks, often with ice or gas, that orbit the sun: Mercury, Venus, Earth, Mars, Jupiter, Saturn, Uranus, and Neptune. [130]

Plant. A living organism that cannot move from place to place, and that produces its own food through a process known as photosynthesis. [53]

Predator. An animal that hunts and eats other animals. [71]

Prey. An animal being hunted or eaten by a predator. [71]

R

Reflection. A bouncing back of light against a surface, such as a mirror. [114]

Refraction. A bending of light when it enters a new material, such as a glass lens. [114]

Renewable Resource. Resources, like trees, that can be replaced or regrown. [158]

Rock. A solid made of minerals found on Earth's crust. [141]

S

Safety Goggles. Special plastic glasses that cover the eyes, used to protect the eye area during scientific experiments. [12]

Seasons. A periodic change in weather caused by the Earth's tilt on its axis as it revolves around the sun: spring, summer, fall and winter. [133]

Sedimentary Rock. Rock made by layers of sand, mud or other materials that become pressed together. Sandstone is an example of sedimentary rock. Fossils are often found in sedimentary rock. [151]

Soil. Mixture of sand, clay, decayed plants and animals (humus) and other materials found on the top layer of Earth's surface. Soil types differ by texture, their ability to hold water, and their ability to support life. [143]

Solar System. The sun and the planets, moons, comets and asteroids that surround it. [128]

Solid. A state of matter in which volume and shape are fixed, such as ice. [92]

Solution. A mixture in which one substance dissolves in another, and seems to disappear in the solution. [100]

Sound. This is caused by vibrating objects. The vibrations carry energy that our ears can hear. Sound always travels in waves through air or an object. [118]

Sun. The star at the center of the solar system. This ball of superheated gases provides most of the Earth's energy. [129]

System. A group of different parts that act as a whole. A system has characteristics that are separate from those of its parts. [53]

T–U

Theory. A "big idea" in science that tries to explain why things happen. For example, scientists came up with the "germ theory" to explain why people became sick. [43]

Thermometer. An instrument used to measure temperatures in degrees. [27]

Tides. The rising and falling of oceans and seas along the shore, usually twice a day. [133]

Tree-growth Rings. The rings created by a new layer of growth around the trunk of a tree. One ring is added each year. Thicker rings indicate more rainfall. [150]

V–Z

Water Cycle. A cycle in which water from lakes and oceans is heated by energy from the sun, evaporates into the atmosphere, condenses into drops of water that float as clouds, and then falls back to Earth as precipitation (rain and snow). [166]

Weather. Conditions in the atmosphere that change daily, such as windy, rainy, or sunny. [164]

Weathering. The gradual wearing down of rocks on Earth's surface by the action of the wind, water, ice and living things. [147]

INDEX